A

创意新古典
CREATIVE NEOCLASSICISM

目录 CONTENTS

中西合璧尊享唯美生活殿堂	006	Chinese and Western Styles Combined Mansion for Wonderful Living
拜占庭艺术绽放万科翡翠别墅	014	A Vake Jade Villa of Byzantine Art
极致华彩，收藏岁月的荣光	024	Luxurious Mansion for Collecting Glory of Days Past
万平拉阔豪宅，见证两百年美国梦想	030	Spacious Mansion: A Witness of Two Hundred Years American Dream
黑白底片上的流金岁月	044	Golden Years Within Black and White Film
蝶舞翩翩芳馥间，月半清风花弄影	052	Butterflies Flying through Fragrance, Flowers Dancing with Half-Moon in Breeze
摩洛哥皇家风范	058	Moroccan Majlis
源于英法的新古典风韵在瑞士绚丽开花	062	The Neo-Classical Prosperity in Switzerland, Rooted in Britain and France
2011CCB	072	Casa Cor BA 2011
定义Neo-Art Deco豪宅新标杆，以艺术为生活加冕	080	To Define Neo-Art Deco Mansion, to Crown Life with Art
浓墨重彩演绎美式风情，谱写海派传奇	088	Presentation of Colorful American Style and Shanghai-Style Legend
丽紫流影，奢尚人生	094	Brilliant Shadow, Luxurious Life
含蓄内敛，铭刻隽永美感	100	Conservative, Connotative and Aesthetic
艺术家居，家居艺术	110	Art's Furnishing, Furnishing's Art
花神之家	118	The Home of Flora
净白素雅，清美出尘	122	White Elegance and Fresh Beauty
波普艺术与美式空间相得益彰	126	Complementarities Between Pop Art and American Space
融古汇今，雕琢隽永之美	132	The Combination of the Ancient and the Modern to Complete Connotative Aesthetics

欧式新古典
EUROPEAN NEOCLASSICISM

源于里维埃拉的灵感，复兴生活真善美	146	Inspiration from Riviera, Truth, Kindness and Beauty by Renaissance
馥满优雅 法式华彩	158	Graceful Elegance, French Brilliance
骑士精神	166	Spirit of Knights
意式底蕴，创造如波托菲诺的旅游天堂	174	Backdrop of Italy, as the Tourism Paradise of Portofino
清新优雅 柔和静美	180	Refreshing, Elegant and Gentle
岁月静好 陌上花开	184	Peace of Time, Flowers in the Field
聚焦惊艳新古典	190	Focus on Neo-Classical Surprise
窗外四季，窗内静水流年	194	Seasons Outside, Peace Inside
雅奢主张之简美风情	202	Minimalism of Sumptuous Grace
净白美宅，静品时光如歌行板	212	The Purity of White to Appreciate a Good Time
碧波上的浪漫	216	Romance Above Waves

美式新古典
AMERICAN NEOCLASSICISM

庄重之美，令人钦佩	232	Solemn to Be Admirable
庄园大宅，豪门气度	236	Manor of Wealthy and Influential Clan
收藏田园梦想	240	Collect Countryside Dreams
天使的空中城堡	246	Angel's Castle in the Air
复刻慢时光	254	Slow Time Once More
意法经典，都市沙龙	262	Italian and French Classic City Salon
醇雅诺丁汉	270	Nottingham of Fashion and Elegance
新手法讲述美国老故事	276	New Way to Tell the Old American Story
360度环景大宅，山谷里的石头居	284	360 Degree Panorama Stone Mansion

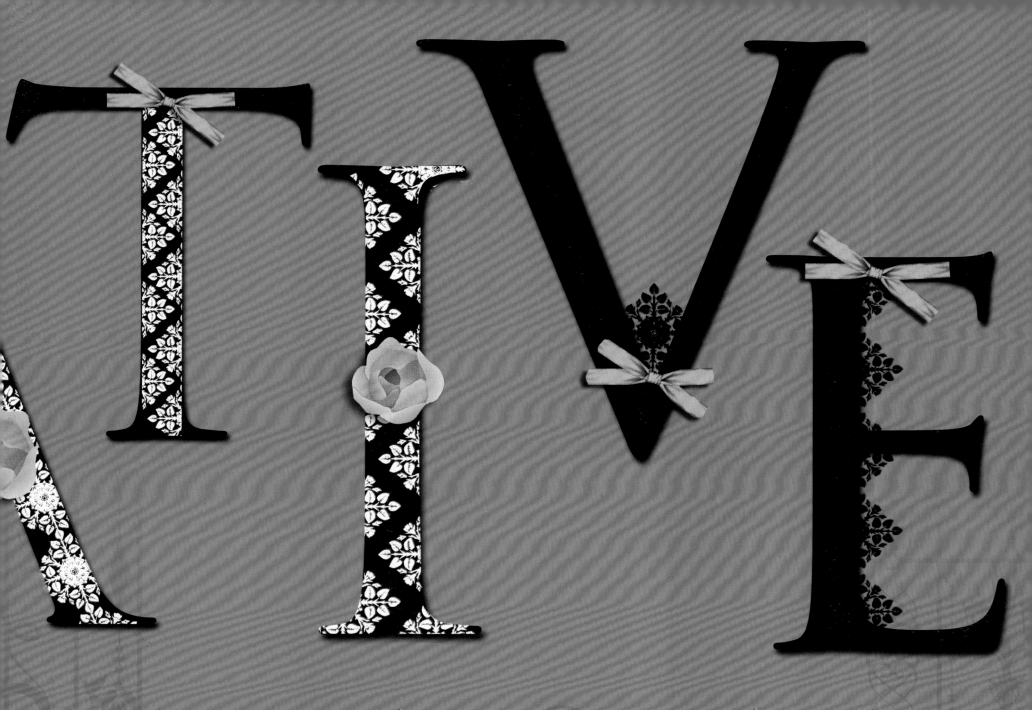

中西合璧
尊享唯美生活殿堂

Chinese and Western Styles Combined
Mansion for Wonderful Living

- 项目名称：成都中海·九号公馆 245 户型样板房
- 设计公司：梓人环境设计有限公司
- 设计师：颜政

- Project Name: Show Flat of 245, No.9 Mansion
- Design Company: Geney Space Design Co., Ltd.
- Designer: Yan Zheng

中海·九号公馆 245 户型样板间的设计是中西文化相互撞击、融合的尝试，它完全颠覆了传统中对法式风格的想象，在拥有强烈的西方设计理念的同时，根据地域特点进行全新的解读，让人叹为观止，不禁发出"原来中西合璧也可以传递出如此完美的法式浪漫"的感叹。

设计将现代的色彩融合中国风古典的纹样，让它们重新组合、演绎和幻化，演变出新的视觉享受，带给人们一种全新的美的体验。以 18 世纪法国最伟大的硬币收藏家约瑟夫·佩尔兰制作的硬币存放柜为参照的柜体，传统的橱柜样式，奇异优雅的漆画以及描金的设计，诠释出现代与经典结合的伟大魅力。而空间中的器物及绘画则可追溯到 18 世纪，深受中国、日本等东方艺术装饰风格的影响。

此外，设计还将一些诗意的内容带进空间的点点滴滴中，如客厅的墙纸，取材于日本江户时代流传到欧洲、美国的绘画，宁静悠远却不失生气的画面，群峦叠峰，却不似巍峨陡峭，用其特有的日式墨画再与中国山水画的深远意境相结合，更准确地诠释出"雅"的别致。

注重灯光与色彩，契合东方人特质

在本案的主卧室里，设计师将一种介于桃红和玫红之间的、名为"蓬皮杜红"的颜色运用其中，烂漫娇艳的色彩，触手成温，引人无限遐想。它比玫瑰的花瓣更加鲜艳欲滴，质感粉嫩，与东方人的肤色非常搭调，让人看起来更美。这种"蓬皮杜红"是法国路易十五国王的情妇蓬皮杜夫人在指导工匠烧制瓷器的过程中偶然发现的，她在艺术、文学和设计领域都有很高的造诣。

当我们置身于卧室里，没有任何防备的时候，肯定希望最亲密的人看到自己最美的样子。于此，设计师通过对色彩、灯光等细节的细腻把控，把这种内心的期许变为了触手可及的现实，恰如其分地表达出人的优雅之美。

正如设计师所期待的，245 户型样板间呈现的绝不仅仅是纯粹的样板间，而是一种居室的生活艺术，一种家庭关系的调节剂。

A design interior this space makes where to see the collision of the eastern and western culture and the trial communication of the two culture, by overturning completely the usual imagination people have towards French style. With a strong concept of the western design, this project makes a new interpretation to achieve stunning marvels according to regional characteristics, exclaiming that the combination of the west and the east can also covey so perfect French romance.

Modern colors is fused with Chinoiserie by reorganizing, interpreting, transforming and making new visual effect for offering an unprecedented aesthetic experience. The coin storing compartment by Joseph, the greatest coin collector in the 18th century in France is referred to for lacquer painting and gliding and copper planting to shape the charm of combing the modern and the classical. Meanwhile, the items and the pictures within can date back to the 18th, conspicuously affected by eastern art deco, like China's and Japan's.

Additionally, something poetic is injected within, like the wallpaper in the living room that is based on pictures started in Edo period, and spread to Europe and America, which are peaceful ad quiet and where there are hills and mountains, though they are not steep or towering. The Japanese ink paining with the far-reaching image of Chinese landscape pictures more accurately brings out the unique of the grace.

More Stress on Lighting and Color, Agreeing with Eastern Traits

In the master bedroom, a color named Pompidou red, more between peach red and rose red, catalyzes endless imagination. It is bright-colored than petals of rose, making a very good match with the skin of

eastern people and boasting people's beauty. Such a color is discovered by Mrs. Pompidou, mistress of Louis XV of France who made great achievements in art, literature and design, when she guided the manufacturing of china.

When in bedroom and with a mood without any preparations made, we are sure to hope our closest person can see the most beautiful of us. So the bedroom sees a delicate control of hue and lighting, turning the inner expectation into reality at hand to flow out the grace and the good quality appropriately.

Just as expected, here presents more life art in a dwelling place, which is a modifier of family relationships, than a pure show flat.

拜占庭艺术绽放 万科翡翠别墅
Vake Jade Villa of Byzantine Art

- 项目名称：万科翡翠别墅
- 项目位置：中国上海
- 设计公司：LSDCASA
- 设计师：葛亚曦、李萍、赵复露、刘德永
- 面积：563 m²

- Project Name: Vanke Jade Villa
- Location: Shanghai, China
- Design Company: LSDCASA
- Designer: Ge Yaxi, Li Ping, Zhao Fulu, Liu Deyong
- Area: 563 m²

翡翠别墅，位于引领百年东西文化交流的上海，由中粮集团和万科集团联合打造，也是上海万科继兰乔圣菲之后推出的又一高端独栋别墅社区。LSDCASA 担纲其软装设计，从价值观和信仰层次入手，着眼当代居住观念及形式诉求，突破风格样式，构筑符合当代上海的居住形式和意识形态审美。

尊贵丰饶的金、冷静睿智的黑、热情灵动的红和正义神圣的蓝，延续着拜占庭艺术色彩美学。家具样式的设计，将拜占庭时期的教堂拱窗、铜灯和繁杂蜿蜒的建筑线条等元素运用其中，成为翡翠别墅的语言。软装方面，设计无视东方与西方、古典与现代的边界，让空间中涌动着拜占庭对差异热烈开放的接纳，对未来的期待和对力量的敬畏。

客厅比较多用到西班牙品牌 Alexandra 的家具，将宫廷的元素反复贯穿于空间，以繁复雕花的铜镜为代表，着力表现拜占庭时期的工艺艺术，而 Maitland-Smit 则将现代款式的家具带入，设计师将古典风格和现代款式的家具结合使用，让空间中活跃着拜占庭的点点精粹，却又不是那么浓烈厚重。

餐厅以宴请为前提，营造就餐仪式，餐桌采用黑色大理石，搭配来自意大利古董豹纹主餐椅，金色 VERSACE 1999 纪念版餐具，构建拜占庭宴会阵列，强调用餐的秩序和礼仪，配合气势瑰丽 Roberto Cavalli 欲望都市系列壁纸，花纹烟熏样式为宴会注入更多生气。

地下一层是家庭活动区，承担阅读和收藏功能，拜占庭艺术被融入到饰品、挂画、家具细节等各处，通过各种样式的组合搭配，形成强烈的视觉符号转换到空间中，营造接近生活又明显高于生活的空间气质。

阅读室 15 世纪华西里·柏拉仁诺教堂复刻的吊灯，世界顶级家具品牌 Cassina 的书架，搭配精美花艺，这里打造的不仅仅是一个家族聚会的场合，更是一个使用者可以尽情取阅的私人图书室，手捧一本书，转身即入戏。

主卧以黑红为主色，搭配金色，营造主卧的不可复制性。以教堂穹顶的图案为艺术创作的挂画，体现拜占庭时期的精神符号。意大利知名面料品牌 LYONTEX 挑战具有历史感的拜占庭风格丝质艺术，带来精致生活艺术的视觉享受。入门处的迷你酒吧则是地下一层私享区的延续，充满收藏意味的小物件和艺术气息的挂画让这个小小的区域更增艺术氛围。

男孩房用大量重复的物件，形成空间的情景化，呼应了男孩热爱音乐的特质。老人房则强调稳重舒适，设计师将拜占庭太阳元素带入空间，同时穿插些许中式意象，演绎出中西交融的视觉美学。

Jade Villa, amounted in Shanghai, a city already returning to a conjunction where to witness the communication of the east and west, makes a high-end freestanding villa by two giant group, including Vanke. Designing completes a dwelling space in line with local aesthetics on the basis of value and belief, and by turning to contemporary living concepts and form, and breaking away the stereotyped style.

Gold, black, red and blue continue the art of Byzantine Art, while furnishings fuse arched window, copper lamp and complicated, winding lines, all typically Byzantine elements. Whether furniture or accessories are free of the regulation of the east and the west, and the classical and the modern to instill the acceptance and tolerance of Byzantine Art, in holding in awe and veneration expectation for future and power.

The living room employs more of Alexander furniture, a name brand from Span, to carry out the elements of imperial palace in reiterated approaches. One example is the bronze mirror with complex carving, exerted to bring out the art exclusive to Byzantine period. And meanwhile, Maitland-Smit introduces the modern ones to refresh the essence of Byzantine style, but not very, by blending the furnishings of the classical and the modern.

The dining room is aimed to set off a ritual sense for dining, where the table is of black marble with antique Italian chairs of leopard print. The VERSACE 1999 table ware is of memorial edition to constitute a Byzantine feast, accentuating the dining order and rite, which with metropolis-series wall paper of Roberto Cavalli and smoked floral pattern inject within more vigor and vitality.

The basement is for family activity, where to house reading and collection with Byzantine Art available in ornaments, hanging picture and furniture. The combination and match of various forms transfers strong visual senses into

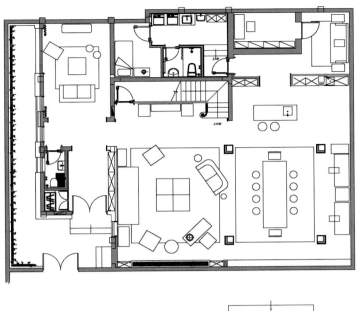
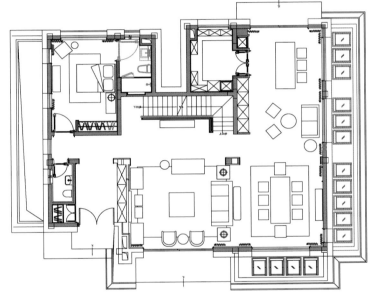
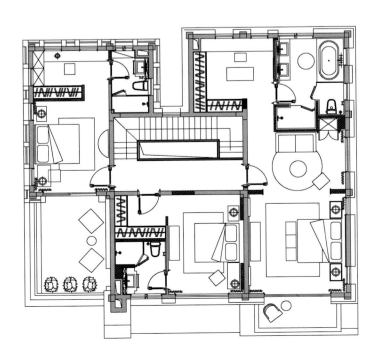

the space to accomplish a temperament closer to but far beyond life.

The reading room features a chandelier duplicated on the one in a church in the 15th century, world-top Cassina book shelf and flower arrangements. Here not only makes a place for family to get together, but more a private library where readers feel nothing but lost in indulging themselves.

The master bedroom is dominated with black and red, as well as interspersing gold to make a unique section. The hanging picture with graphic inspired with the vaulted roof in church to embody the spiritual remark of Byzantine era. LYONTEX, an Italian famous fabric poses a challenge to the Byzantine silk art, bringing visual enjoyment of delicate life and art. The mini bar at the entrance continues the style of the basement with connotative and meaningful items and hanging picture to boast the artistic ambiance.

In boy's room, objects appear repeatedly and in large amounts to present scenes to the space, echoing with personal enthusiasm of and passion for music. As for the room for parents, it attaches more significance on staidness, calm and comfort, where to take in solar elements of Byzantine time with Chinese images to make a mixed aesthetics of the west and the east.

极致华彩，收藏岁月的荣光
Luxurious Mansion for Collecting Glory of Days Past

- 项目名称：海岸别墅
- 设计公司：基里尔室内设计与装饰
- 设计师：基里尔
- 摄影师：米哈伊尔

- Project Name: Portofino Villa
- Design Company: Kirill Istomin Interior Design & Decoration
- Designer: Kirill Istomin
- Photographer: Mikhail Stepanov

"海岸别墅"顾名思义与海岸咫尺之隔。如此得天独厚的条件决定了空间的内部设计。轻松的气氛、明亮的细节、精彩的贝壳石装饰给人一种度假般的感觉。整个空间无地毯之铺设，但大部分家具却如同穿上了轮滑一般。轻盈的质感背景映衬着设计的严谨。家具铺陈或古色古香，或定制。天花的贝壳、母贝造型由法国艺术家量身打造，历时10个月。水晶吊灯虽着上了金属的外衣，但却给人以"珊瑚"的意象。吊灯、水晶、海贝的灵感源于20世纪美国设计大师托尼·杜奎特。蜈蚣般的长椅、木质的粉色孔雀石桌面都是本案设计师的杰作。虽然很多物品的制作由多个欧洲工作室承担，但包括在欧洲海运，在美国购买的众多事宜皆由本案设计完成。

别墅可谓是花园洋房，庭院深深，典型的当地风格。空间公私分明。隐私区除了卧室，还有一个小小的生活区。公共区除了大客厅，还有方便聚会的客厅。光彩照明一边是人工照明留下的轻盈质感，一边是天然的太阳光线，让空间有了一种透亮、无色的感觉。

The location of this villa, just steps away from the coastline, determined its decor. The interior is light as vacation mood, with bright details and fancy rocaille forms. There are no carpets in the house, and most of the furniture is 'dressed' in slips. Although it looks so lightweight, everything is serious here in terms of interior design. Every single item here is either vintage/antique or custom-made. For example, it took 10 months for French artisans to create the overhead moulding of seashells and mother-of-pearl. Same for the fairytale chandelier with metallic 'corals', rock crystals and seashells that was inspired by famous American 20th century designer Tony Duquette. The furniture is made after Istomin's fantasy design - centipede benches and wooden tabletops of non-existent pink malachite. While some objects were under construction in various European ateliers, Istomin was hunting for everything maritime on European and American auctions. The catch looks impressive - for example, Dorothy Draper's commode.

Villa's square construction with a patio inside is typical for that region. After redesign it was divided into two parts - private (with a small living room and bedrooms) and ceremonial with grand dining room and gala living room. The light is phenomenal here - what looks outrageously bright in other places here, under the sun, seems pale and colorless.

万平拉阔豪宅，见证两百年美国梦想
Spacious Mansion: A Witness of Two Hundred Years American Dream

- 项目名称：萨里山山庄
- 项目位置：英国萨里郡
- 设计公司：伦敦 MPD
- 设计师：毛里齐奥
- 建筑公司：好房子乡村别墅
- 摄影：杰克摄影

- Project Name: Surrey Hills Country
- Location: Godalming, Surrey, UK
- Design Company: MPD London
- Designer: Maurizio Pellizzoni
- Architect Company: Fine Town & Country House Commissions Ltd.
- Photography: Jake Fitzjones Photography

"萨里山山庄"的历史可以追溯到1901年，爱德华七世时代的地产讲究艺术与工艺。曾经的建筑空间，新加了一个双高的正式厅——橘园漠漠，主卧多了三个卫生间，还有一个步入式衣柜。主人旧时的收藏，海外旅行时的所得共同容纳于同一空间，给人一种折中但却个性化的美学。重新设计的厨房是主人与当地专业厨房制作者的杰作。卡拉拉的工作台自然是意大利的品牌。细部的精心考究让设计有了一种高端的优雅。国际大牌的家具、面料运用于整个空间，让奢华有了一种极富品位的精致感。

一万平方米的超大空间升级更新，由MPD与建筑师协力打造。高大上的"美国化"糅合着业主的私藏古董与艺术品。为了整体的协调，所有的艺品都得到了重新规划。包括厨房用具、走道扶手、沙发等在内，20%的物品都由MPD定制。每一个卧室在设计师的构思下，都得到了命题。内部空间，如客厅自在舒适，两个户外阳台也是充满了惬意之感。

Dating from 1901, this large Edwardian property is rooted in the Arts and Crafts movement. Working closely with the architect to ensure space was used to maximum effect, MPD extended the original building to add a new double-height formal room, Orangery, three extra bathrooms and walk-in wardrobe for the master bedroom. The client was keen to incorporate existing pieces of furniture from various family trips abroad into the new design. These pieces were distributed amongst the rooms in the house, and used as the inspiration behind each room's individual color scheme and feel, creating an eclectic yet personal overall design aesthetic. MPD completely redesigned the kitchen alongside a local expert kitchen maker, sourcing the perfect finishing touches like the custom made Carrara worktop for the kitchen. Particular attention was paid to such final details throughout the property in order to achieve a high-level of elegance and sophistication in the design. Ralph Lauren and Andrew Martin furniture and fabrics were also used throughout to create an air of luxurious refinement, while complementing the client's existing pieces.

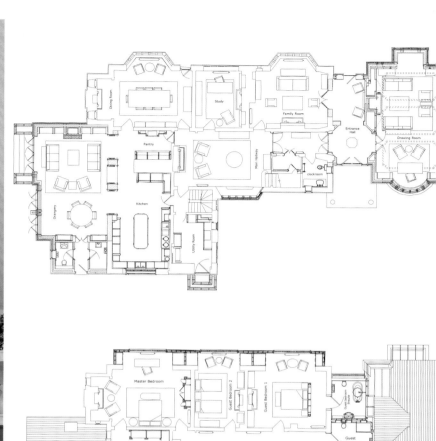
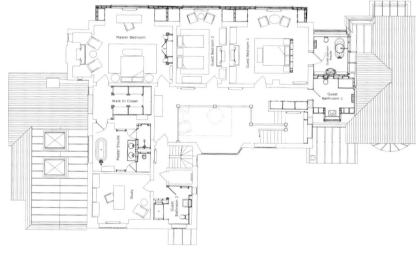
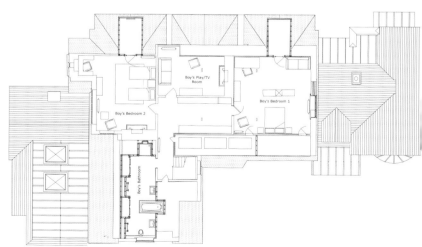

MPD London collaborated with an architect to renovate from scratch this 10,000 square meters family mansion in the Home Counties. The brief was to recreate The Hamptons in the Surrey Hills, and Maurizio set about creating an upscale 'American look' that incorporated the owners' many personal antiques and objets d'art. All the artworks were reframed for consistency and around 20 percent of the furniture, including the kitchen, the balustrade in the hallway and all the sofas, was custom-made by MPD. Maurizio themes each bedroom around an existing antique piece, and extended the effortless comfort of the living rooms to two outdoor terraces.

黑白底片上的流金岁月
Golden Years Within Black and White Film

- 项目名称：圣热尔曼公寓
- 设计公司：安多工作室
- 建筑师：阿齐扎
- 设计师：安娜、埃蒂

- Project Name: Sait Germain Apartment
- Design Company: Ando Studio
- Architect: Aziza
- Designer: Epstein Anna, Azougy Ety

本案位于巴黎。333平方米的空间，临街而立。业主希望本案能以全新的面貌、电脑化的设计充分挖掘空间潜力，以便起到促销的目的。

本案设计流程包括一系列的3D细节测试，包括家具摆放、墙面的涂饰、木材等建材的使用、地板的风格、艺术、配件等在内，都颇费思量。本案设计周期约为2个月。对于法式精髓、细节，设计尽可能予以保留。同时，设计融清新、现代于整体空间。家具铺陈全部由米兰最好的大师设计，并由知名公司制作。

如此设计的结果，自然是法式的细部考究，现代、清爽的线条。黑的幽暗、亮的光泽，现代的配件、古董等器物得以平衡于同一空间。在此居住是如此富有个性、有趣。

The apartment is located in Paris France. The apartment is 333 square meters and its facing the street. The client is a private owner of the apartment and he hired our services since he wanted to sell the apartment and by our design and computer visualization of the design he could show the full potential of the interior spaces.

The design process included a lot of 3D tests of details, positions of furniture, finish materials such as walls colors, woods, flooring styles etc. It also included a lot of work with the client for choosing the right furniture, art and accessories. The design process lasted 2 months approx. The design was trying to keep the spirit of the French style and details and yet give it a fresh and modern twist. All furniture was chosen from best Milan designers and cutting edge well-known companies.

The final result is a very classic apartment keeping all molding French details yet using very clean and modern lines. The balance between dark and bright, modern accessories and antique artifacts. All this made the apartment very unique and interesting place to live in.

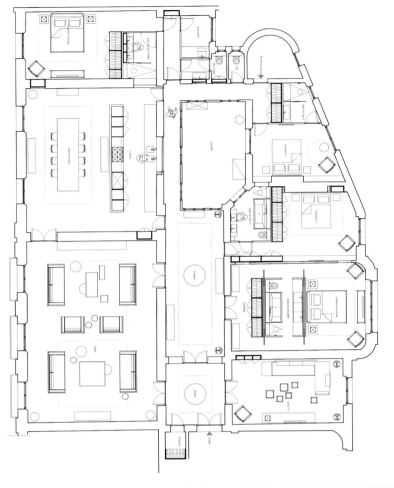

蝶舞翩翩芳馥间，月半清风花弄影

Butterflies Flying Through Fragrance, Flowers Dancing with Half-Moon in Breeze

- 项目名称：昆明中航·云玺大宅"玺悦墅"中式户型
- 设计公司：深圳市则灵文化艺术有限公司
- 设计师：罗玉立
- 用材：桦木、金属、综合材料
- 面积：873 m²

- Project Name: Seal-Pleasure Villa, Air China (Kunming)
- Design Company: Zest Art
- Designer: Luo Yuli
- Materials: Birch, Metal, Comprehensive Material
- Area: 873 m²

本案别墅装修在中式风格中融入西方元素，整体空间用色颇具新意，以蓝色为主色调，点缀以金色，同时元素运用也很恰当，运用少量的金属材质突显时代感。设计用线条打造出方与圆的相融，每一细节的精心装饰，让整个空间显得乱中有序。天花板的木线条拼贴、暗红的实木地板，再加上实木家具的装饰，使中式的古典、时尚的装修设计溢满空间。

设计用吊顶分割出空间，中式镂空屏风隔断，延伸了空间，并且改变以往传统的红色调，增添时尚元素。白色的百叶窗装修设计，斑斓的阳光映射进客厅。每一处都有花草的装修设计，每一个细节都恰到好处。宴会空间蓝与白的相融，描绘出云南的特色，又显现出宫殿式的装修效果。

中式风格讲究对称，橱柜的镂空"回"字纹，不失时尚感的白色调，别有韵味的红木椅，精心的雕刻，时刻呈现传统元素的新时尚。用浅蓝色圆点装饰背景墙，落地窗的白色纱幔，与当地气候环境融为一体。一幅山水画，艺术的插花，修饰了空间。壁橱之间的精巧小柜绝对是亮点，明黄的颜色，丰富了空间层次。

古典的宫廷椅沙发，精致的雕花，造型各异的抱枕，让小空间显得不那么单调。浅灰色的背景墙没有过多的装饰，用栩栩如生的模型演绎大雁南飞的盛况，在这里品茗，欣赏古典的艺术美，感受天空的开阔和自由。而利用画的造型设计的隐形门，以及层次不同的手绘画等，既美观实用，又丰富了空间的层次。

高贵典雅的女孩房，简洁的床品，用飞舞的蝴蝶装饰自由美。床头的背景墙，以米色的小方格为背景，照片墙的设计配合两盏开放的花灯，一切都是那么自然简单。床边的楼房造型装饰品与墙面上的壁纸两相呼应，对面的飘窗装修设计，华贵的公主幔帐，打造欧式风格的浪漫经典。而男孩房墙面上的钟表设计是一个亮点，用颜色打破空间单调的装修格局，丰富空间的层次。

负二层平面布置图

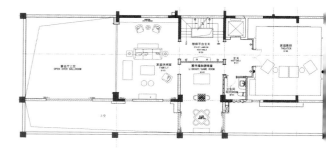

负一层平面布置图

一层平面布置图

二层平面布置图

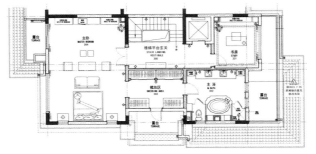

三层平面布置图

A project this space makes that blends western elements into Chinese style, where the pigments are innovative with blue master and golden servant. Elements are used very properly and appropriately with small amounts of metal materials to highlight the sense of era. Lines accomplish the harmony of circle and square. Each detailed part is decorated elaborately, so the space on the whole seems messy but actually quite orderly. The parqueted ceiling of wooden bar, the solid wood flooring of dark red and the furnishings and accessories of Chinese style overspread the classical Chinese everywhere.

Suspended ceiling works now as the partition. Hollowed-out Chinese screen lengthens the space in separating. Traditional red undertakes alternation with fashion element added. Through the white louvers comes patches of sunlight. All witness the appearance

of flowers and plants. Each and every detail is done to the very point. The blue and the white in the feast space portrays features of Yunnan, while embodying a decorative effect of imperial palace.

Chinese style as usual stresses the symmetry. Cupboards are hollowed out with fret, white yet of fashion sense. The unique chair of rosewood and the meticulous carving bring out the latest fashion out of the tradition. The white gauze curtain down the landing window acclimates the local surroundings. A landscape painting and the flower arrangement catch time to ornament the space. The subtle locker between the cupboard absolutely makes bright spot, with its brilliant yellow to enrich the spatial hierarchy.

Thanks to the court-couch sofa, the carving and the cushions, small sections is no longer monotonous and dull. With overmuch ornament, the grayish backdrop with models makes a vivid scene where wild gooses fly south. Here is a place for tea drinking and classical art appreciation to feel the broad of the sky and the personal freedom. The backdoor in form of picture and hand-painted drawing of different layers are both practical and beneficial to enrich the space.

The room for the daughter is of noble elegance, where simple bedding features dancing butterflies, the backdrop of bedhead is set against small beige diamond, and the photo wall well cooperates with two flower lamps. All look so natural and simple. The decoration of building model beside bed echoes with the wallpaper. The bay window and the mantle for princess, altogether accomplish a European romance. As for boy's room, the clock design is really a light spot, which with hues breaks away the bald layout while enriching the space.

摩洛哥皇家风范
Moroccan Majlis

■ 项目名称：摩洛哥皇家风范
■ 设计师：扎伊尔 (www.mtaher.net)

■ Project Name: Moroccan Majlis
■ Designer: Muhammad Taher (www.mtaher.net)

清爽的线条，极简的感觉，是优雅、安静而无丝毫令人烦恼的干扰。美与优雅却依然得到了释放，如一朵盛开的花。帝王宫殿般的内部设计真如专为王公权贵而设计，而不是为了那些寻求噱头的"广告狂人"。

宫殿应有的标志本案入口早已经彰显。即便庆祝只是偶尔举行，这里也有足够的座位与大量的花艺。"颓废"在这里似乎成了正当的理由，也有了适当的借口。难怪这里到处都有皇室的影子，原来是西班牙巴伦西亚侯爵府带来的大量理念。

客厅里，每个角落都有着软装考究的沙发。中央处一个巨大的桌子不管是夫妇共同进餐，还是晚间举行派对彰显都是舒适。紫色，皇室的御用色，如今于本案空间得到了大量的运用。奢华的绒面床头，照耀睡床的别致水晶灯，让卧室除了舒适还是舒适。

甚至走廊都可感受到皇室的标志。织锦的墙纸与镜面升华空间的开阔与宏伟，即便等待，在这里也是享受。

With its clean lines and minimalist feel, it is elegant and soothing without being too fussy. The images that are showcased here, however, are the opposite of this subdued aesthetic. The CG artists featured have created regal interiors better suited for royalty than a character on Mad Men.

This grand entryway features all the trappings of a palace. With plenty of seating for the occasional gala and massive flower arrangements, this is truly a space for the independently wealthy with a flair for decadence. Artist Veltik took a good deal of inspiration for this space from the Palacio del Marques de Dos Aguas in Valencia, Spain.

This sitting room, with carefully upholstered sofas in every corner and a massive table in the center is the perfect place for a couple to sneak away, or for an entire party to continue the evening in comfort. Purple has long been a color associated with royalty, so the fact that artist M Taher uses it here is no surprise. From a luxurious tufted headboard to a unique crystal lighting feature over the bed, this room screams decadence and comfort.

Even hallways can feel regal with the right trappings. Here, the use of a brocade wallpaper and mirrors make the space feel big and expensive, so no one could complain about waiting here for a few moments.

源于英法的新古典风韵在瑞士绚丽开花
The Neo-Classical Prosperity in Switzerland, Rooted in Britain and France

- 项目名称：瑞士别墅
- 设计公司：基里尔设计与装饰
- 设计师：基里尔
- 摄影师：米哈伊尔

- Project Name: Swiss Chateau
- Design Company: Kirill Istomin Interior Design & Decoration
- Designer: Kirill Istomin
- Photographer: Mikhail Stepanov

"瑞士别墅"位于日内瓦，风景秀丽的滨湖地带，一派湖光山色，好不舒适。建筑量体灵感源于英国建筑，并有法式风格之元素。基里尔负责二楼的室内设计，包括主卧、卫生间、化妆室、私人会客厅、客人房及儿女房。折中的风格下，彰显着悠悠的民族风。众多的细节，极其考究。大量的奢华建材、家具铺陈、古董及定制的器物闪烁于其中。卧室、客厅的墙面。古色古香的软装面料，给人一种久远的装饰感。

到处都是定制元素的身影，无一点大一统的痕迹。墙体的面料装饰、铜灯是20世纪30年代的身影，地毯是40年代的风情。空间另外彰显着一个主打元素，那便是金色系的运用。很多器物镀金镏光，为的是整体的整齐划一。女主人专用卫生间，装饰着水晶柱、铂金架。铂金柱上涂构的紫水晶色调，更是把奢华挥发到了极致。

For the bedroom quarters of this Swiss chateau, situated on the picturesque lake of Geneva, opulence and comfort were paramount. The building is inspired by traditional English architecture with elements of French style. Istomin was in charge of the second floor with master bedroom, bathroom and dressing rooms, private living room, guest bedrooms and two kids rooms – for the son and the daughter. Although the interior looks eclectic, it is dominated by the classical style. This project contains an enormous number of details. Istomin has used a wide range of expensive materials and furniture pieces, lots of antiques, replicas of the museum pieces and custom-made items. For example, bedroom and living room walls are upholstered with an antique fabric that recreates ancient ornaments.

This interior is fully custom-made – you can hardly find a catalogue pieces here. Fabrics on the wall and bronze lamps are copies of the 1930s items. The carpet ornament is copied from a 1940s species. Another common element is golden shade. Many pieces were gilded so that they are of the same shade. Her bathroom has the most expensive trimming – columns of crystal, platinum racks with amethyst handles.

2011CCB
Casa Cor BA 2011

- 项目名称：2011CCB
- 设计公司：SQ 建筑师事务所
- 设计师：法夫里希奥
- 用材：大理石、原木、丝绸、地毯
- 面积：178 m²

- Project Name: Casa Cor BA 2011
- Design Company: SQ+ Arquitetos Associados
- Designer: Fabricio Scarcelli
- Materials: Botticino Marble, Rosso Verona Marble, Wood, Silk Fabrics, Pastel-Colored Classic Rugs
- Area: 178 m²

设计以早就存在的空间器物为基础，也因此成就了设计与环境的平衡。众多的物品，每一件都必须立意明确，彼此间要和谐共处，对于设计是个挑战。

为了突显许多的器物，人造灯不得不精心安排。但同时，自然光线同样不得忽视。因此，白天的空间一如平常，平淡而不显眼；夜晚下的空间，却充满了动感。众多的开口，通风的同时，沟通着内外之间，如同一个天然的空调，极好地应对着当地的高温。除了卧室、卫生间因为隐私的需要，其他各处功能空间多为组合设计。不同时代于此交汇，经典与现代在此融合。现代的家居生活，空间质性却非常贴近于殖民时代的风格。

舒适的气氛，文化的感觉，多元的质感，旧时年代重视家人及社交生活的气氛无不弥漫于整个空间。所有配件在与空间文化质感相互融合的同时，年代虽各有所属，但彼此间却相互尊重。偌大的一个空间，无缝对接地实现着穿越。

经典成就着舒适与奢华。家具、器物、艺品各展个性的同时，和美地存在于本案空间。

The art objects existed before the spaces were designed; so, the decoration arose from them. Thus, the collection embodies the eclecticism impressed upon the environment, which is plenty of stories to be told. The major challenge was to establish a dialog between so many important pieces, not allowing each one to obscure the other, and to achieve the creation of a classical harmony with so disparate objects.

The artificial lighting was meticulously planned in order to enhance the works of art; however, there is enough natural lighting in the space, which produces two versions of the environments, one during the day (more usual) and another one during the night (more dramatic). Regarding ventilation, the environments have openings that provide a good amount of natural air flow, and have also the benefit of an air-conditioning system which serves all environments, being this residence in Salvador, where average temperatures are very high. The spaces are integrated, exception made to the bedroom and WC, due to the need of some more privacy. So, we proposed a mixture of two ages, classical and contemporary, in which the spaces' distribution (usage) happens in a current manner, while the aesthetic composition is closer to the classical (colonial style).

075

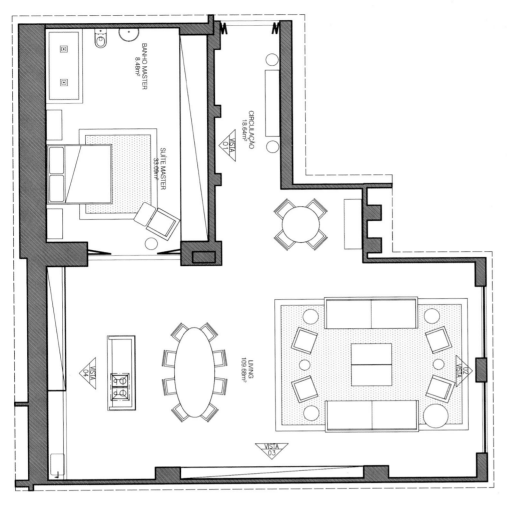

What attracted me the most in this project is the atmosphere we got, i.e. the comfort and culture feeling created by the diversity, along with the rescue of a time in which people valued their families' and society's history. All accessories were chosen in a manner to provide the spaces with culture, always respecting each one's relevance for its respective time period. By doing so, we seamlessly walked through several periods, from the XVIIth century to present days.

The project's conception was based on the classical period, from whose style we sought to rescue the comfort and luxury by providing the environment with furniture, objects and works of art which are representative of such a style.

定义 Neo-Art Deco 豪宅新标杆，以艺术为生活加冕
To Define Neo-Art Deco Mansion, to Crown Life with Art

- 项目名称：财富公馆·御河城堡
- 项目位置：北京朝阳区
- 软装设计：LSDCASA
- 设计师：葛亚曦、蒋文蔚
- 面积：1 600 m²

- Project Name: Fortune Mansion & Imperial River Castle
- Location: Chaoyang District, Beijing
- Upholstering Design: LSDCASA
- Designer: Ge Yaxi, Jiang Wenwei
- Area: 1,600 m²

中国十大别墅之一——财富公馆·御河城堡地处北京朝阳区，是财富地产集团为名门贵胄打造的领航产品，被誉为中国顶级别墅代表。

该别墅样板间五年前由知名设计师邱德光先生设计，因其大胆张扬的风格在当时被认为是邱先生新装饰主义风格的代表作。中国设计思潮发展非常迅速，无论是观念还是意识形态，人们对设计的要求不再停留在某设计风格的样式表层，从单一的追随转向思考和面对自身的独特需要，注重挖掘能对应精神认同层面的需求和设计本身应有的人文关怀。财富集团为了再造中国顶级新装饰主义豪宅，特别委托 LSDCASA 担纲财富公馆·御河城堡的别墅样板间改造，期待展现设计特点的同时，再次代表财富阶层的生活。

LSDCASA，延续建筑及室内的新装饰主义风格为基础环境，续写丰沛的美学力量空间，以匹配财富阶层应有的生活方式，当今世界，科技及生活的发展为人们提供了各种程序和解决方案，无谓追求单一的设计审美，让多元化争论和质疑互为存在，设计在本案中再次发挥创新的力量，打破既有程式，让单一的权力、财富的显性诉求过渡到生活中对伦理、礼序、欢愉、温暖的需要。

这套 1 600 平方米的府邸共有三层，从地下一层逐步向上，空间的每一层都有自己独特的功能和对应的趣味与隐喻。

门厅，室内设计的调整配合建筑空间倾向于表现特质与规律的设计意图，保留近似公共空间的尺度之于人的压迫感的力量，把墙面常有的明显后现代徽章图案的墙纸更换为有层次稳定力量的色彩，配合意大利基于传统审美却又藐视规则、大众习俗的设计和装饰艺术，让空间拥有近似庙宇或会议厅般的神秘庄重。

客厅贯穿门厅的设计风格，设计师在家具陈列上采用了强烈的对称和仪式感，色彩是这里最大的礼赞，以冷艳高贵的钻石蓝与沉稳大气的咖啡色为色彩基调，搭配璀璨的金色和经典的黑白休闲色调，从天花到四周，从家具到靠垫，从饰品到绿植，无不展现了待客空间的华贵。

西餐厅以沉着的墨绿色调为主色调，搭配浅绿的窗帘帷幔，点缀蓝色与白色的精致花艺，餐厅一隅，雪白的孔雀拖着一袭长尾妆点着华美的空间，让这座中西交流的空间层次起伏，生机盎然，鸟语花香。独具风格的中餐厅则兼容了大户宴客排场和文人精神，餐厅古典实木家具，精致黑白插画的屏风，辅以餐厅中精美的花艺，糅合出平衡典雅的用餐氛围。

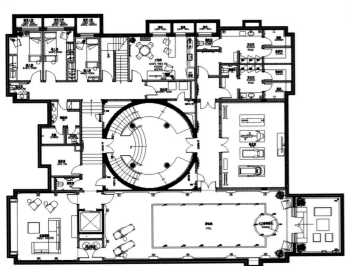
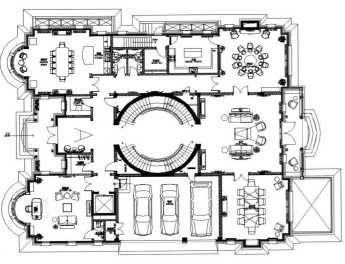
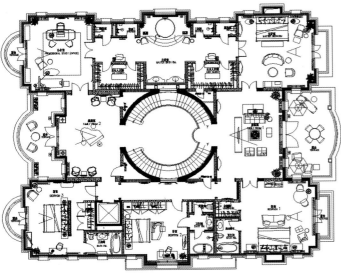

走上二层，可以看到一个个色彩平衡、层次丰富的卧室空间。米色和咖啡色系是这里最经典的色彩基调。在这个基调上，设计师融入不同层次的紫色和蓝色，一会轻快、一会沉稳，为不同的主人营造韵味十足的私密空间。

书房里，是主人收藏与展示的空间，各种藏品和摆件演绎格调，通过从细节到整体的微妙处理，男主人温文尔雅的外表之下，对品质生活的追求得到了完美的诠释。地下一层作为主人娱乐和休闲的区域，分布着恒温泳池和影音室，纵观整套邸府，更像是具备魅力和非凡感官的艺术臻品，时光就此凝练成艺术，生活由此完美升华。

This project, as one of Top 10 villas, is a leading project by its owner and honored as the high-end representative among the same level in China. Five years ago, the sample house of the real estate was done by Qiu Deguang, a famous designer, whose bold style was thought a magnum opus of neo art deco by Mr. Qiu. In the past years, design in China has undertaken a rapid development, where design requirements have shifted from the single follow-up to personal demand instead of the skin layer of a style by stressing the spiritual respondence and the humanistic care inborn with the design. And now, in order to present a top mansion of new art deco in China, the refurbishment of the sample house ever by Qiu is aimed to bring out features of design and life of the wealthy elites under the hands of LSDCASA.

On the basis of continuing the art deco and the spatial aesthetics, LSDCASA makes innovative efforts to break away the stereotyped practice and transmit the monotonous resort of power and wealth into needs of ethic, ritual, happiness and warmth. All are met for matching the lifestyle of the rich group, not seeking dull design for people with multi choices in a modern setting of science and technology, and taking plural arguments and questions under the same roof.

Of the 3 floors of 1,600 square meters, each is endowed with its own function, interest and metaphors.

In the foyer, the interior coordinative to the building is intended for presentation of a unique style. The tension similar to that by public space is kept; the post-modern wallpaper is replaced with hue layers that's stead and powerful. Design and art deco in line with aesthetic tradition but defying rules allow the space

with a mysterious solemnness alike to that of temple and conference hall.

The same style is carried out throughout the living room, where furnishings and accessories are rich in symmetry and ritual sense. Color makes the lightest spot. The tone of indigo and brown is interspersed with gold, black and white. From the ceiling to the horizontal setting, all items like furniture, cushion, ornament and greenery confides in nothing but luxury.

As for the western dining room, its hue is blackish green, which is accompanied with curtains and mantles of light green, as well as delicate flowers of blue and white. In one corner, a snow-white peacock makes a pose with its long tail. The Chinese dining room is inclusive of requirements of rich and influential families to hold banquet and literati's spirit. The classical furniture of solid wood and the black-white inset screen with floriculture complete a balanced and elegant atmosphere.

The 2nd floor is world of balanced colors and rich layers. Beige and brown are the most classical, into which are fused purple and blue of different layers to build up private space with a lasting appeal, though sometimes light and then staid.

The study is for personal collection and display, where articles from the detailed to the holistic make a perfect interpretation of the quality life the button-down host pursues.

The underground is used for entertainment and leisure to accommodate swimming pool of constant temperature and audio-vision room. The whole space is rather like an art piece with special charm and extraordinary organ impact, where time is developed into art and life is then sublimed perfectly.

浓墨重彩演绎美式风情，谱写海派传奇
Presentation of Colorful American Style and Shanghai-Style Legend

- 项目名称：长滩一号
- 设计公司：十上设计
- 设计师：陈辉
- 摄影师：周跃东
- 撰文：李芳洲

- Project Name: No.1 Long Beach
- Design Company: Tenup Design
- Designer: Chen Hui
- Photographer: Zhou Yuedong
- Text: Li Fangzhou

每个人对家的追求总是孜孜不倦的，希望通过自己的努力实现梦想中的生活。与此同时，人的心境也总随着年龄的增长而不断变化，越发沉稳，喜爱的格调也渐渐往简单、沉静靠近。少了分年少轻狂，多了分成熟魅力，而家也以一种内敛的方式，来诠释一种生活理念。

在这个居家空间里，"木"是最鲜明的主角。木质的门、木质的橱柜、木制的装饰，这些枣棕色的木料沉淀了家居的氛围，让空间大气十足。空间色彩以暖色调为主，围绕着木色进行过渡，深与浅的搭配，明与暗的融合，形成丰富的空间效果，置身于其中有一种安然与自在。整个居家的格调虽选用美式风格，却在细节中大胆地采用混搭手法，让空间融入一丝活力与趣味。公共空间的地面选用复古砖，以特色的拼贴方式拼凑出有趣的效果。在客厅沙发的位置处，抛弃了常规的地毯而以瓷砖拼成地毯形式，更贴近空间的氛围。

本案位于江畔，拥有得天独厚的美景。设计师将原有同客厅相连的内包式阳台打造成一个小型休息区，并打造了一个榻榻米。天花用条木塑造出屋顶样貌，闲暇时坐于此看看书，喝喝茶，欣赏江景是何等惬意。它与客厅间呈开放式的布局，使得空间感觉更为宽敞。客厅以一组墨绿色的沙发吸引人们的目光，搭配上改良的中式坐凳以及中式挂画、矮柜……这些中式的元素映衬在美式的大背景下，中与西的结合，耐人寻味。餐厅与客厅毗邻，间隔处用四色玻璃打造的隔断，透过玻璃看餐厅，每一种色彩都能创造不一样的空间"味道"。隔断同时也起到景墙效果，造景手法在这里挥洒自如，悬挂的鸟笼装饰，大型的花卉盆栽，描绘出一幅动人画卷。

设计师通过细节的处理，规避了走道狭长、压抑的景象。不论是门上细致的卷草装饰，还是悬挂在天花之上的星星灯，或是在拐角处的孔雀玻璃装饰，处处隐藏着惊喜。卧室的布置则简洁明了，只保留了基本的功能装饰。高大的床、明晰的线条和优雅的装饰，充满自由随意的舒适。

通过生活的积累与对艺术的喜好，设计师塑造了这个独一无二的家居空间。木的纹理在不同角度，不同灯光下产生不同光感，使得整个居家充满文化气息与历史韵味，而家也回归纯粹，更加耐看。

Everyone is diligent in pursuit of a dwelling space with an aim to realize life in dream with own efforts. Meanwhile as time goes by, man is up to be grow mature and increasingly staid, more preferring something simple and calm, and no longer frivolous. A home space simultaneously is beginning to interpret a life concept in a reserved approach.

And in this home, wood is conspicuously the protagonist. Door, kitchen cabinet, and decoration are all of wood. Such a jujube-brown material is lost in a dwelling atmosphere, space thereby becoming magnificent. Centered on warm hue, the space make a wood-oriented transition, dark with light, light with bright to make rich spatial effect for people to have feelings safely and free. Against the dominant American style, details employs mix-match approaches to allow for vigor and fun. The flooring of the public space is of retro brick, specially parqueted to make interesting effect. Around the sofa in the living room, the usual carpet has been abandoned by using tile to piece carpet image, much closer to the spatial air.

The location along the river bank is incomparable to have more views. The inbuilt balcony attached to the living room is shifted into a tatami room, where the ceiling is of barwood to shape a roof image, and nothing gives more pleasure than reading, tea drinking and river scape appreciation. Both of the tatami room and the living room are designed open to make the space more spacious and wider. The very green sofa in the living room is more eye-catching, which with improved Chinese benches and Chinese hanging picture and low cabinet are worth afterthought against an American setting. The boundary between the living room and the dining room

dome with glass of four colors. Eyesight through each color glass gets different view. The partition of glass also works as a landscape wall. With free landscaping approaches, the hanging bird cage and the large flower pot culture trace out a vivid picture. Treatment in details offsets the narrowness and pressure of aisle. Whether the grass rolling above on the door, or the star lamp down the ceiling, or peacock glass decoration reveals surprise everywhere. The arrangement in the bedroom is simple, with basic functional decoration kept. The tall bed with clear lines and elegant decoration is filled with comfort free and optional. The understanding of life and the preference of art shapes here a unique home. At different perspectives, wood generates more light sensation with light cast on to make the home have more cultural air and historical sense. And home returns to purity, standing careful reading and appreciation.

Brilliant Shadow, Luxurious Life
丽紫流影，奢尚人生

- 项目名称：常州金新鼎邦新古典风格别墅
- 设计公司：G&K 设计
- Project Name: Jin Xin Ding Bang Villa
- Design Company: G&K Design

本案别墅豪宅将服装设计前沿的紫色大胆运用到家居生活空间中，设计师让高贵典雅与时尚奢华成为该空间设计的真实表达。藤制的户外沙发安放在宽敞的露台中，不失为阳光午后，朋友小聚的理想场所。

设计师运用紫色与芥黄的对比冲击力，使客厅空间色彩冷暖相辅相成。大面积的紫色系布艺沙发和紫色高档布艺定制绣花窗幔配合暖色压花地毯，彰显出干练、优雅、时尚、奢华的韵味。

紫色缎面的餐椅在水晶灯的照耀下，让别墅餐厅表现出一种绚丽剔透的优雅。开放式厨房墙面采用了紫色马赛克造型拼图，与岛台紫色皮艺的高脚餐椅相得益彰。采用透明玻璃贴合木饰面的高地收藏展示柜位于地下室展示收藏区的中心区域，错落有致、低调奢华。

The edge-leading purple used in fashion is boldly used in this villa, where the combination of the noble elegance and luxurious fashion make the expression in its real sense. Sofa of rattan is fixed onto the terrace, making an ideal place for children in the afternoon.

The contrast between purple and mustard yellow makes the cold and warm hues in the living room complimentary. Fabric sofa coated in purple of large area and the expensive custom embroidered curtain of purple coordinate with the warm knurling carpet, embodying a taste refined, graceful, fashionable and luxurious.

The dining chair wrapped in purple satin brings out a brilliant grace of the whole space, once with light cast on. Walls in the open kitchen are embellished with purple mosaic, setting off the best in each other with the purple leather of the island. The display cabinets with transparent glass and wood veneer in the core of the basement display area are well arranged high and low, low-key but still sumptuous.

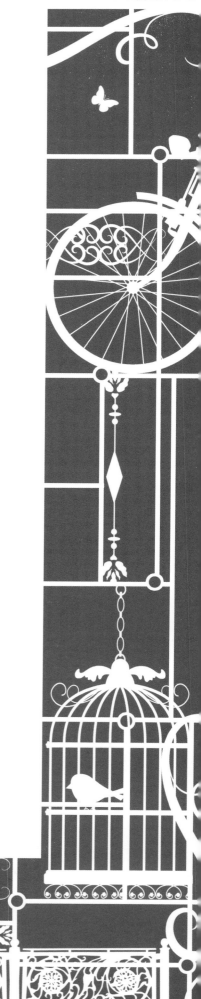

含蓄内敛，铭刻隽永美感
Conservative, Connotative and Aesthetic

- 项目名称：东莞晓庐假日田园样板房中联排 B1F~3F
- 项目位置：广东东莞
- 设计公司：天坊室内计划
- 设计师：张清平
- 面积：251 m²

- Project Name: Show Townhouse of B1F~3F
- Location: Dongguan, Guangdong
- Design Company: Tian Fun Interior Planning Ltd.
- Designer: Zhang Qingping
- Area: 251 m²

低调奢华在本案有了全新的诠释。住宅应该像一块璞玉、一瓶陈年美酒，暖暖内含光、越沉越香醇，经得起时间的考验，居于其中越久越能发现其美好，而唯有内敛不张扬的美感，卓越精致的工艺方能达到此目的。以点、线、面精巧布局，把这座家宅当成雕塑品般雕琢，也像铺陈画作。看似简单的设计里，其实蕴藏了以时间雕琢的精致、以艺术质感铺陈的奢华，而这些都藏在细节中，需要居者慢慢去体会。

为了让空间量体就像一件雕塑品，从空间贯穿到家具，例如以线条雕琢的天花线板、大理石地板、墙垣，当代经典、线条简洁的沙发，直线倾泄宛如银丝瀑布的水晶灯，精巧的多宝格开放式储柜，直线拼花的艺术性地毯，而除了线条的铺展之外，能让这件空间雕塑散发光芒的还有材质的讲究、光线的运用、品牌沙发的皮革质感、艺术品的细腻质地等。在此把家具也视为空间的延伸，让彼此融合为一体，透过一些家具突显艺术性，创造戏剧张力与视觉效果，让空间更有生命力。

Here makes a refreshing interpretation for reserved luxury. A dwelling space should be an uncut jade, or a bottle of old wine. The better a jade is, the more illustrative it looks, The more ancient a wine is, the more fragrant it smells. For a house, the more you live within, the more good qualities you will find. Such a house, however, needs excellent and delicate process.

The layout of point, line and surface make the space like a sculpture, or a painting overspread. A simple design actually contains delicacy done by time and luxury with artistic texture. But all are in details and takes the occupants to appreciate.

In order to make the space more like works of sculpture, lines are used to string the space and the furnishings, like those for ceiling, marble flooring, wall, and sofa. Additionally, the chandelier looks like waterfall with silk lines dropping downward, the open cabinet to collect more small items is of delicate lines, and the art carpet is of line parquet. Besides lines, the careful use of material and the effort paid to lighting boast the glister of the spatial sculpture, like the leather of brand-name sofa, and the delicate texture of art pieces. Furnishings then are viewed as the extension of the space, to integrate each other. The artistry of some furniture is set off to create dramatic and visual effect, so that the space is filled with more vigor and vitality.

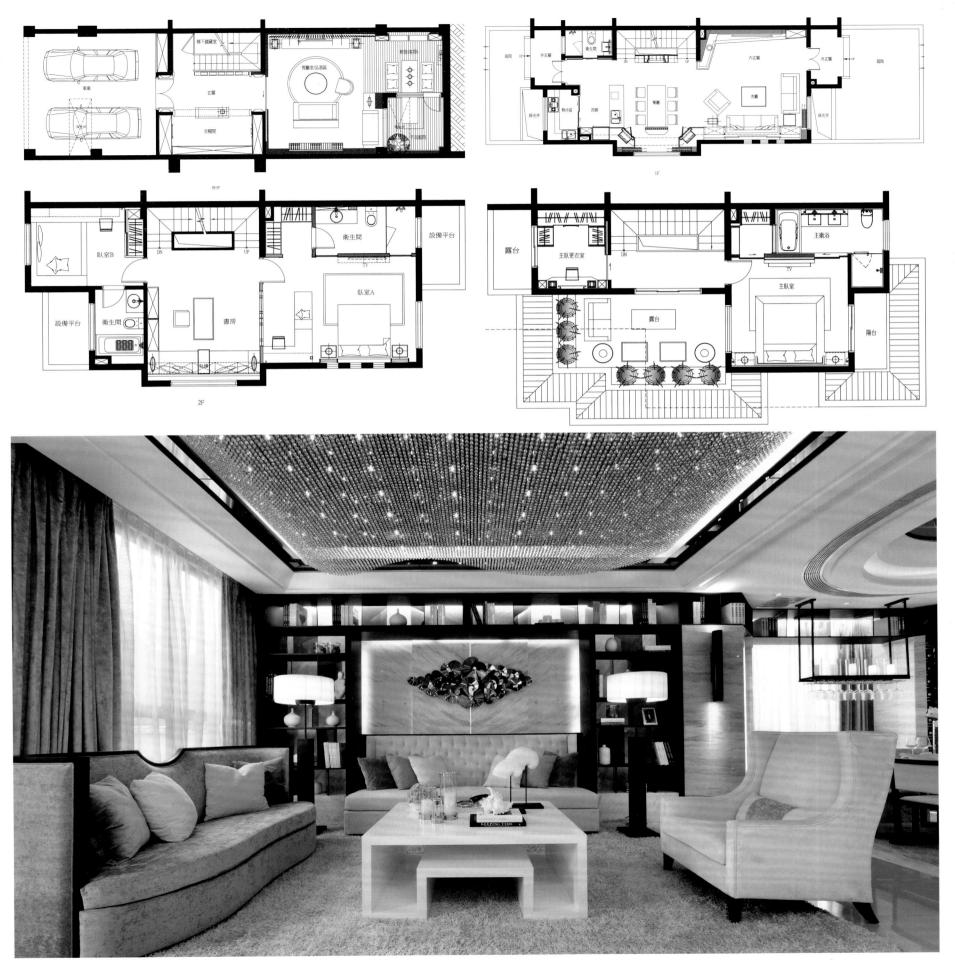

艺术家居，家居艺术
Art's Furnishing, Furnishing's Art

- 项目名称：迷藏
- 项目位置：中国国际会展中心
- 设计师：孟可欣
- 参与设计：刘洋
- 摄影师：史云峰
- 用材：家具、罗奇堡家具、西班牙乐家洁具、亚威护墙板、意大利 Zanaboni 家具
- 面积：1 000 m²

- Project Name: Hide and Seek
- Location: China International Convention Center
- Designer: Meng Kexin
- Participant: Liu Yang
- Photographer: Shi Yunfeng
- Materials: Furniture, Sanitary Fittings, Chair Rail
- Area: 1,000 m²

在邻近公司的位置设计师根据客户的需求，构思一个多功能会所，体现家和艺术的双重概念，前半部分为艺术品展示，后半部分则作为豪宅生活样板展示。在一个连贯长廊式 L 形空间中分出三种不同不连贯的风格，以此来满足不同客人的需求。

记得勒·柯布西耶说过，百年后人们会谈论一种"风格"，但今天只有一种风格，即一种精神格调。它蕴含于作品，蕴含于真正的创作之中。是的，就是精神格调！设计的目的是格调，而格调的来源却是人。本案展现了三种不同格调的人，禅者、优雅者、权贵者。

禅者——宁静的力量

禅是在"空"中构筑"有"，用无形来表达有形。正如法顶禅师说过，放下越多的人，得到的越多。正因为放下，才会有力量，一种迟达的力量。正如雕塑家野口勇所说，对石头过分加工，它就很难以成为雕塑，石头也会死去。雕塑是以形的形式出现的，重要的是它完成之前的思考过程。

入口的雕塑是艺术家王瑞林的作品《迷藏》——大圣入定相，齐天大圣的神通广大与坐禅入定神妙的结合，巧妙地讲述了力量与宁静的主题。洽谈区的人体座椅来自意大利设计师法比奥·努维伯尔（Fabio Novembre），而桌子则是定制老榆木。两种不同性格的家具混搭，让空间充满戏剧性色彩。如果说人体座椅是后现代戏谑的表现，那么桌子则是自然的表达。而黑色和原木色看起来深沉宁静，却在人体的曲线雕刻中暗含着某种不安的力量。

优雅者——融化一切色彩

优雅是和谐，是松弛，是曲线，是柔缓，是艺术的产物。他们从来不跌跌撞撞，永远都表现得圆融而舒缓。原则上优雅者从不拒绝任何色彩，却能融化一切色彩，融化得那样有诗意且自然而不留痕迹。他们是造物的高手，能与神对话的使者。

开阔的大厅划分为三个区域，两组围合沙发区域和一个洽谈区。色彩运用则是经过反复推敲的。背景的米黄色护墙，蓝色的沙发，酒红色的座椅，不同比例的摆放着，体现了所有的原色构成，视觉上饱满丰富。同时，象征着法式的古典时尚，酒红的波尔多，米黄的阿维尼翁城堡无不融入其中。家具款式上采用了法

国品牌RocheBobois，优雅时尚，而空间用途更多体现了巴黎的某种沙龙文化。

权贵者——沉稳和奢华

你可以看到他们的界限，那样不可触碰。有时甚至有些夸张，但不可否认权贵者有自己构造天堂的方式。雕塑式的表达方式，珍贵稀缺的物料，精雕细琢的手工制作，超越人体的尺度，无不诠释着自己的空间性格。而此刻权贵与美瞬间凝固了，一种交响乐式的表达。

古堡式的就餐长桌，托举人物大理石雕刻壁炉，科林斯浮雕柱式样胡桃木护墙，拿破仑帝国风格的单人沙发，意大利巴洛克风格的家具，早期文艺复兴"十"字图案的实木拼花地板。这些元素交织在一起相互凝视、呼应、共鸣。

The project is near the company the owner works with. Personal requirements contribute to a multi-functional chamber to integrate dwelling and art into one. Its form part is used for art display while the rear for displaying mansion life. a continuous L-shaped space presents three successive styles to satisfying demands of the guest.

"In one hundred years, people will talk about one style, but today sees only one. That is the spiritual level, which is implied in works that is accomplished by real creation", said Le Corbusier. Sure, that is the spiritual level. Design is for style, while style originates from people, ones of three kinds, one of Zen, one of grace, and one of power.

The One of Zen: Power of Peace

Zen is aim to build up physical existence in emptiness, with the invisible to make the visible. According to Boep Joeng, the more you give up, the more you get, because only when you put down, you have more power. Isamu Noguchi, a sculptor, says, overmuch process is exerted onto a stone, and the stone would in no way become a sculpture, because an exhausted stone can die away. Though sculpture always appear in form, but the more important is the thought and idea to be applied into the process before.

The sculpture of *Hide and Seek* is done by Wang Ruilin. Mahatma is sitting quietly and meditating. The far-reaching supernatural power of the Great Sage Equal of Heaven is combined with the meditation, telling of the theme of power and peace. The human body chair in the communication area is accomplished by Fabio Novembre, a designer from Italy. The table is custom and of elm. The materials of two different properties are mixed together. If the chair is a joking expression against the current, then the table is natural. The black and wood, though looking tranquil, implies some uncertain strength in the curves of human body.

The One of Grace: to Melt All Colors

Grace is about harmony, relaxation, curve and softness and the result never stumble, for they are always harmonious and relaxing. A man never deny any hue by principle, for they are able to melt all colors without any trace. They are experts at creating, and can be envoy to s with God.

The lobby consists of three sections: sofa area of two groups of sofa communication area. The hue use is tested in computer. The beige

the blue sofa and the wine red chairs are arranged in different ratio. All are made up of primary colors to have a solid vision. Meanwhile, the French classical fashion, the wine red Bordeaux, the beige Avignon castle are also available. Furnishings of RocheBobois, a French brand is involved throughout. The space on the whole is to embody some Salon culture of Paris.

The One of Power: Staidness and Luxury

You can see the boundary of these powerful persons, which can by no means be broken off even to an exaggerated point. What's not deniable is that these influential officers always have their own way to build up their paradise. And the sculptural expression is to treat rare materials with meticulous approaches with an aim to go beyond size of human body for interpreting the spatial traits. It's right here that the power and the aesthetics are melted in a frozen state, making a great symphony.

The dining table in is castle form; the fireplace is carved with marble; the walnut parapet wall is of Corinth relief column; the armchair is of Napoleon style; the furnishing is of Baroque style; and the cross-shaped parquet flooring of solid wood seems to go back to the early Renaissance. All are interwoven and echoing with each other.

T 花神之家
The Home of Flora

■ 设计公司：城市设计 ■ Design Company: Chains Interior
■ 设计师：陈连武 ■ Designer: Chen Lianwu

Flora，在画家 Botticelli 的画作《春》里，是花神的名字，身着白色洋装的花神，衣服上画满了各式的花卉植物，充满生机盎然与春天百花齐放的气息。由于女主人 Flora 对于花卉植物非常有兴趣，也极有研究，本案便以此发展，将花卉植物的元素，带入各个空间。

空间以白色为基底，将白色作为画布，利用不同颜色与图案的壁纸，搭配家具及窗帘，堆栈出各空间的层次与氛围，营造出不同的自然风情。

Flora, the goddess of flowers, is dressed in white dress with various flowers and plants in Botticelli's famous painting *Allegory of Spring*, full of vigor and vitality exclusive to spring. The through study into flower of the hostess leads to the elements of flower and plant available in every room.

White is the base color for the space and as a white canvas. Different colors and wall papers are used to match furniture and curtain, creating numerous nature textures and multiple layers in each section.

净白素雅，清美出尘
White Elegance and Fresh Beauty

- 项目名称：白格子
- 设计师：伊万
- 摄影师：费尔南多

- Project Name: Classic in White Lattice
- Designer: Iwan Sastrawiguna
- Photographer: Fernando Gomulya

"白格子"的展示空间及书橱以贝壳石壁龛作为装饰。壁龛之下有石质的镜框围塑着壁炉的形象。餐厅与走道之间的墙体上部装饰着白色的石格子。

玄关哥特式，石膏拱顶天花。此处除了四个石膏立柱，还有隐式的向上投射的照明及铁质的吊灯。

厨房岛柜覆盖有马赛克，环绕般地对接着"L"形状的黑色花岗岩台面。该台面的石膏托梁专业定制，可以容纳五人同时就餐，也是家人早餐时的最爱。

玄关内有核桃木的圆桌，有插花，中央处还有吊灯。整个空间以暗色的木器，银色的长方形镜面，锈铁的烛台作为空间的背景。咖啡色的天花细部，彰显着华丽的石膏造型。

Stone seashell niche used for display and bookcase with stone mirror frame and faux fireplace mantel below. White stone lattices are placed on top of the walls between dining room and hallway.

Gothic FOYER with plastered arched ceiling, four plastered columns with hidden uplights, and iron chandelier.

Kitchen Island is covered with tumbled stone mosaic and is attached with L-shape wrap around black granite counter top. Plastered corbel is custom made to support the L-shape counter top. This kitchen island has become the family's favorite breakfast nook. Five stools are provided below it for everyday use.

Walnut round table with flower arrangement and iron chandelier are placed in the center of this foyer. Bow front dark stained wooden chest with silver leafed rectangular mirror and rustic iron sconces sit as a background. Coffered ceiling is embellished with ornate plaster molding details.

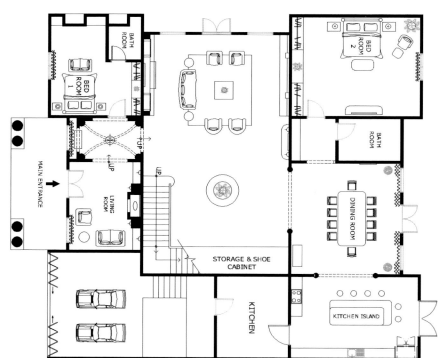

波普艺术与美式空间相得益彰
Complementarities Between Pop Art and American Space

- 项目名称：上海绿地凯旋宫联排别墅
- 软装设计：LSDCASA
- 设计师：葛亚曦、李萍
- 面积：300 ㎡

- Project Name: Townhouse of Triumph Palace, Greenland (Shanghai)
- Upholstering Design: LSDCASA
- Designer: Ge Yaxi, Li Ping
- Area: 300 ㎡

该项目位于上海松江区，整体设计冷静时尚、优雅且颇具国际范儿。设计师希望从文化品位、生活礼仪、审美情趣的范畴建立认同与产生共鸣。空间的主要色调为黑白与暖咖，祖母绿色的点缀比较巧妙，有时是抱枕、沙发，有时是挂画、花器，甚至是空间中一些细腻的小细节，可见设计师之用心。波普元素的大胆融入，让整体空间在现代美式的格调下更显前卫与时尚。再加上精致的家具与摆设，材质对比中和谐统一，空间呈现出一种成熟的美，百看不厌。

Located in Songjiang District, Shanghai, calm but fashionable on the whole, the project is graceful and very global, which is aimed to make resonance from cultural taste, life ritual, aesthetic sense and identification. Against the black, white and coffee, the green of mother pearl works to embellish skillfully. Even the cushions on the sofa, or the hanging picture, the flower container, or even trifle detail provides a glimpse, from which we can see small but delicate details. Pop elements are fused boldly, so the whole space looks more fashionable and smarter in a modern American setting. Together with the delicate furnishings and accessories, and the materials contrasting harmony, the space becomes more mature, at which you never feel bored and tired even you have already looked at it over 100 times.

融古汇今，雕琢隽永之美
The Combination of the Ancient and the Modern to Complete Connotative Aesthetics

■ 项目名称：台北文华东方酒店　　　　■ Project Name: Mandarin Oriental, Taibei

台北文华东方酒店坐落于人文荟萃、绿意盎然的台北市商业中心敦化北路上。酒店拥有 256 间客房与 47 间套房，品位卓绝的客房至少 17 坪（55 平方米）起，以古典雅致融合当代风格的设计理念，提供全台北市最宽敞舒适的住宿空间。所有客房皆备有最新颖的科技设备、豪华大理石浴室与可容人进出的宽大衣帽化妆间。气派豪华的总统套房及文华套房各占地约 114 坪（400 平方米），备有专属的 Spa 与健身空间，让宾客享有极致奢华的尊宠与舒适；此外，酒店享负盛名的东方汇（Oriental Club）提供商务旅客更贴心专属的住宿礼遇。

台北文华东方酒店的餐厅和酒吧以隽永细致的创新视觉美学为概念，精心设计出独具风格的时尚餐饮空间，提供国内外消费者顶级奢华的美食飨宴。国际知名设计大师季裕棠为酒店量身打造三间风格截然不同的餐厅，现代时尚的法式小酒馆式餐厅（French Brasserie）"COCO"提供经典法式餐点和亚洲美馔；高雅的粤式餐厅"雅阁"提供定制化服务，让宾客在优雅的环境中品尝精致美食，并备有数间包厢供私人宴客使用；由米其林星级意大利主厨 Mario Cittadini 领军的意式餐厅"Bencotto"，透过开放式厨房，宾客可尽兴享受意大利主厨们在愉悦氛围中精心制作的地道意大利家传风味佳肴。

除了餐厅外，宾客可在位于酒店优雅中庭的"文华饼房"，品尝由世界级巧克力大师 Frank Haasnoot 亲手制作的欧式西点、蛋糕与手工巧克力糕点；"MO Bar"则提供多样化的调酒与香槟供宾客选择；酒店大厅的"青隅"（The Jade Lounge）更是台北都会中享用文华东方著名精致下午茶点的最佳去处。

酒店内的芳疗中心引进集团屡获殊荣的芳疗概念与闻名遐迩的疗程，提供结合健康、美容和按摩的完整体验。宽敞的芳疗中心占地两个楼层，共提供 12 间芳疗室，其中包含 4 间双人芳疗室和 2 间 VIP 套房。设备齐全的健身中心与 20 米长的户外温水泳池，满足宾客在台北都会生活中恣意享受轻松悠闲时光及各项休闲健身的需求。

Amounted on Dunhua North Avenue, a place gathering talents and sheltered in tree shades in Taipei, the hotel of Mandarin Oriental offers the most spacious accommodation throughout the city with its 256 rooms and 47 suites. Even the smallest guest room reaches an area over 55 square meters, where to combine modern design with the classical grace. All are equipped with the state-of-the-art science and technology, marble bathroom and walk-in cloakroom. President suite and Mandarin Oriental suite each cover an area of 400 square meters, have their own Spa and exclusive fitting room, for guests to enjoy the utmost dignity and comfort. Additionally, the Oriental Club offers the most considerate and thoughtful accommodation.

Its restaurant and bar are endowed with meaningful and meticulous innovative visual aesthetics and concept to shape a modern catering space with unique style and top-level feast across the world. Ji Yutang, a world-renowned designer, offers three distinguished exclusive restaurants: French Brasserie where to provide modern and fashion French pub, COCO offering food in refined ambiance and Fine Pavilion where to entertain guests with custom service in a good setting. Boxes serve well for people to have banquet. Bencotto, conducted by Mario Cittadini, a Michelin-star chef from Italy. The open kitchen gives guests authentic flavor and food handed down from generation to the next to their heart content.

Besides the dining room, the bakery in the atrium presents European cakes, sweets and hand-made chocolate pastry. All are done by Frank Haasnoot, a famous chocolate master in the world. In MO Bar, there are various mixed drinks and champignon. The Jade Lounge in one corner of the lobby makes an ideal place to have afternoon tea in Taipei.

The awarded fragrance care and courses of treatment are available in the Fragrance Center, where to have a complete series of health, skin care and massage. On two floors, the Center has 12 rooms in all, four being double rooms and two VIP suites. The facilitated fitting center is fixed with a 20-meter hot swimming pool outside, where to have relaxation, entertainment and exercise.

PEAN

NEOCLASSICISM
欧式新古典

源于里维埃拉的灵感，复兴生活真善美

Inspiration from Riviera, Truth, Kindness and Beauty by Renaissance

- 项目名称：苏州九龙仓国宾一号N520别墅样板房
- 设计公司：上海邂舸装饰设计工程有限公司
- 设计师：David Desmond、朱凯华、杨良冬、梁瑞雪
- 面积：788 m²

- Project Name: N520 Villa
- Design Company: Shanghai Ya Ge Decoration Design Engineering Co., Ltd.
- Designer: David Desmond, Zhu Kaihua, Yang Liangdong, Liang Ruixue
- Area: 788 m²

此项目室内设计风格是基于法国里维埃拉地区的别墅风格，以此只是为了唤醒隶属于蒙特卡洛的奢华和暧昧。它具有迷人法式阳台、高达十米的水晶吊灯、处于中庭的旋转楼梯和楼梯栏杆上无法复制的鎏金图案，以及只出现在法式宗教建筑中的拱形石材柱廊和带有名贵宝蓝石图案大理石地面等，这一切的初衷只是为了实现项目定位：出入皆人物。

传统的欧式室内风格中并存着多种元素，大量繁琐沉闷的细节充斥其中，随着国内设计界的不断成熟和国际优秀设计作品的涌入，我们对于所谓欧式风格有了重新的认识，让人们不再苦寻"繁琐至上"的居住理念，而是希望能从生活出发，回归建筑功能的本质——居住。在这个项目中，本案运用了大量的雅士白大理石，并通过空间秩序的分割和引入大量传统建筑单体细部元素的排比为别墅原始空间进行了大量优化。

一层的挑空客厅和餐厅的连接采用了拱形柱廊的元素，使两个并不连通的空间，从视觉上融为一体，再从整体装饰颜色配搭上区分两个空间功能上的不同，餐厅使用热烈华贵的红色，并辅以金色元素点缀，让仪式感较强的长桌就餐区域看起来不至于太过于正式和严肃，客厅的挑空中四面墙的元素均较为生动。高耸的壁炉壁画和同二楼连通的卧室回廊都是为整个空间添光增彩的小动作。一楼次卧及书房的设计考虑到室外花园的绿意盎然，所以在室内色系的选配上考虑使用了部分绿色。父母房不再是重色的专属领地，也可以随意的将其定位为子女的居住空间。地下室的概念主题是诺曼底号船舱，有一定的Art Deco风格，旁边的区域分别设置了两个空间——麻将室和女主人活动区，主题较为雅致，从而削弱了由船舱主题所带来的工业金属感。

当真实被唤醒，一切善与美才能复兴。

This is a project which is inspired by the villa style in Riviera, France, aimed to arouse the luxury and ambiguity exclusive to the Monte Carlo. Here has French balcony, chandelier measuring as high as ten meters, spiral stairs in atrium, the gliding pattern that cannot be duplicated on stair railing, which with arched stone portico and marble flooring with sapphire blue pattern are to realize the position of this project: all are human-oriented. Among the European interior style, there are numerous elements, with floods of complicated but stiff details. As projects at home grow increasingly mature and more good qualified are rushed from abroad, designers for this project have refreshed their understanding of the European style. Rich people don't seek a dwelling concept of living utmost, but hope to restore building nature into living. So in this project, large amounts of white marble are used, cut in line with the spatial order, and meanwhile, more details of traditional building are employed to optimize the originally messy space.

The emptied living room and the dining room on the 1st floor involve element of arched colonnade, visually linking the two ever-separate spaces. Meanwhile, two functional spaces are differentiated with hues to decorate on the whole. The dining room is of red with gold available occasionally, the area long table of ritual sense looking not too formal or solemn. Elements on the four walls in the dining room are all life-like. The towering fresco around the fireplace and the corridor of the bedroom attached to the 2nd floor are adding luster to the brilliance. The secondary bedroom and the study on the 1st floor are landscape-oriented beyond the garden, so the green in some parts is a natural choice. The room for the parents is no long a manor exclusive to heavy colors, which is sustainable and flexible to be shifted into rooms for children. The basement accomplished with a theme of Normandy cabin, is of Art Deco. Two spaces are fixed of mahjong room and hostess activity. The graceful and elegant theme grounds off the sense of industrial metal by cabin theme.

Only when the truth is awaken, can the good and the beauty be restored.

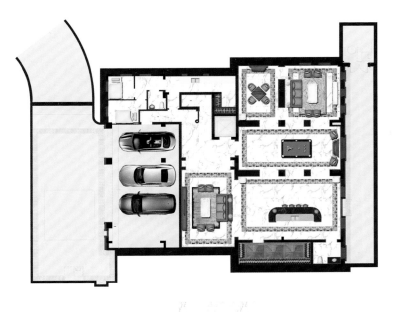
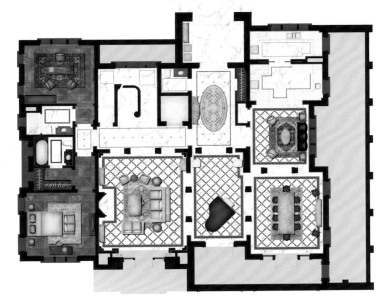

G 馥满优雅 法式华彩
Graceful Elegance, French Brilliance

- 项目名称：成都中铁奥威尔洋房样板房
- 项目位置：四川成都
- 设计公司：尚辰设计
- 面积：165 m²

- Project Name: Garden Villa of CREC (Chengdu)
- Location: Chengdu, Sichuan
- Design Company: Sunson Design
- Area: 165 m²

本案是位于成都中铁奥威尔洋房的样板房，设计师以轻松自然、温馨浪漫为主题，以纯洁的白色和淡淡的浅蓝色为基调，通过完美简练的新古典线条、优雅的蓝调软装配饰以及精益求精的细节处理，打造一场让人惊艳的视觉盛宴。整个空间跳脱出豪宅的固化样貌，多了一份清新、自然的韵味。

起居室中造型新颖的欧式线条沙发、鸟笼状搁置架、茶几、衣柜、儿童床等皆采用白漆木材质，大方、简洁的线条为空间平添一份富有质感的清新美，在适当的饰物点缀之下，丝毫不显得空旷，反而给人一种宁静感。蓝白主色大面积展现于花园阳台软装上，晶莹剔透的纯白波帘尽显轻盈质感，创造梦幻般的仙境。米色贵妃榻上鸟类彩绘抱枕与艳丽的牡丹插花为整个素净的空间增添自然生机。

本案每个房间都有自己的主题，各种细节极富生活情趣，试图营造一种归家的浪漫与一种精致的小资情节，从而让她爱上这个家。主卧起居室延续蓝白主色调，湖蓝印花壁纸墙面与原点彩面三折屏风，加上暗绿波帘，整个空间尽显淡雅、清新之气。线条优美的湖蓝色贵妃榻与墙面芭蕾舞者油画图隔空呼应，为整个空间植入极富艺术性的美学概念；渐变色条纹壁纸与纺纱波帘营造出一个通透、明快的室内休憩空间。

空间中随处可见的花朵元素，各种动物造型和工艺品，既能起到调节室内空气、低碳环保的作用，同时也点亮了新古典家装的自然气氛，犹如在无拘无束的大自然中，自由、惬意。

A show flat this project is for a garden vial, which takes ease, nature, warm and romance as its theme. Against the tone of white purity and light blue, neo-classical lines impeccable and simple, graceful blue upholstering and details kept improving all together make a visual feast. Away from the luxurious stereotype, the whole space has a taste refreshing and natural.

The modeling-innovative European lines of sofa, the bird-cage-like shelf, the tea table, the wardrobe, the child's bed are all of painted-white wood, generous and simple to allow for aesthetics refreshing and rich in texture, which with proper accessories looks anything but empty, but more peaceful and quiet. The white and blue coating the upholstering on the balcony and the wave curtain of white create a fairy land. The painted cushion on the couch and the flower arrangement of the peony add more vigor and vitality to the simple space.

Each room has its specific theme, displaying numerous details of life with an aim of shaping romance returning home and delicacy to stimulate the love for the home out of the inner heart of the occupants. The master bedroom continues the white and blue. The printed lake-blue wall paper, the three-face screen and the very green wave curtain, make the space look more elegant and refreshing. The couch with beautiful lines and oil painting of ballet dancer echo with each other across the space, injecting the artistic philosophy into the space. The stripped wallpaper with its color changes slowly with the gauze wave curtain builds up a clear and crisp resting area.

Floral elements everywhere, animal images and art ware not only meditate the air in the interior, low-carbon and lighting up the natural sense of the neo-classical furnishings, where you seem to feel free and leisure like in nature.

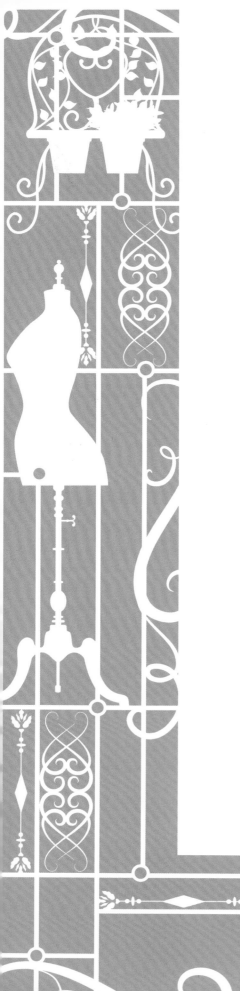

S 骑士精神
pirit of Knights

- 项目名称：融科地产橡树澜湾别墅样板房
- 设计公司：重庆品辰设计
- 设计师：庞一飞、颜飞
- 用材：石材、做旧显纹漆、线条描金处理、手绘油画、实木地板
- 面积：320 m²

- Project Name: Show Flat of Oak Bay Villa
- Design Company: Chongqing Pinchen Design
- Designer: Pang Yifei, Yan Fei
- Materials: Marble, Paint, Gliding Line, Oil Painting, Solid Wood Flooring
- Area: 320 m²

本案汲取法式风格精粹，大手笔制作，体现出恢宏的气势及豪华舒适的居住空间，呈现出中西交融的华贵与浪漫。在布局上采用对称，细节处理上大量运用雕花、线条描金处理，同时在空间中辅以雕刻精细的家具，崇尚冲突之美。

整个空间秉持典型的法式风格搭配原则，空间均为米白色配深色系，表面略带雕花，配合扶手和栏杆的弧形曲度，显得优雅矜贵。随着时间的流逝越发富有历史感的古雅气质，成为镌刻岁月的永恒空间。由人而家，由家而生活，空间的每一个布局，都应体现生活的态度。品辰设计把整栋别墅户型的潜力充分发掘出来，用空间语言诉说奢华而又精致的生活态度。

从陈设上看，体现出欧洲上层社会的贵族文化——骑士精神。骑士有其骄傲的一面，还有谦卑的一面。谦逊的态度不仅仅是面对年轻貌美的女士和身份显赫的贵族，在对待平民时，骑士也绝不会恶言相向，体现出精致的奢华。一场空间盛宴，不止宏大，在装饰细节中完全秉承了古典主义的精美庄重，各种曲线更是细节之处的灵魂。在静态空间中流淌出优雅，仿佛穿越数世纪的浪漫时光。

品辰设计，源自生活，源自内心。上善之品，浑然天成。

With essential quality of French style, this is a dwelling space majestic and magnificent, and of luxury and comfort, where to offer brilliance and romance of the east and west. Symmetrical in layout, the project features carving, gliding lines as well as delicate furnishings to make a sharp but intended contrast.

Typical of French style, the holistic interior is wrapped in off-white, with carving to collaborate with handrail and banister to allow for an unconstrained elegance. As time goes by, the space looks more historical but eternal. A space it is done with ideas that people determines their house, and the house gives birth to their life. Every inch of the space should be reflection of life. The possible potential of the whole villa is completely realized with spatial vocabulary to tell of an attitude of extravagance and delicacy.

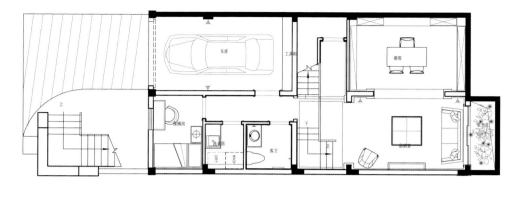
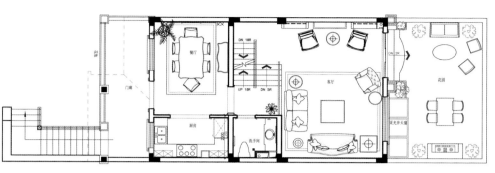

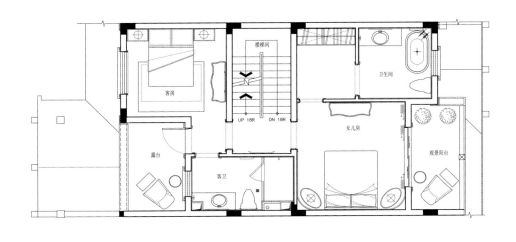
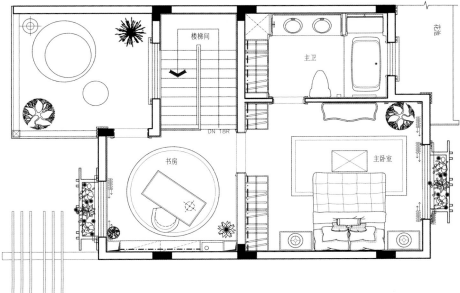

Furnishings embody the spirit of knights, a culture of upper class. Knights are both proud and humble, who exert the same courtesy to young woman, distinguished noble man and ordinary people. A feast in a space should not just be grand, but embody a solemn beauty of classicalism. Curves are the soul of details, flowing out grace and elegance to make romance that feels as if it has gone through dozens of centuries.

Here really makes a project rooted in life and inner heart, one of utmost quality that seems to have been done by Mother Nature.

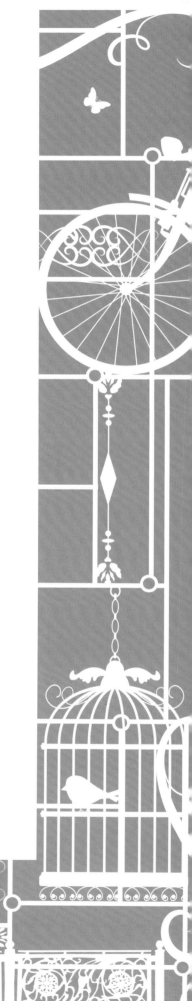

意式底蕴，创造如波托菲诺的旅游天堂
Backdrop of Italy, as the Tourism Paradise of Portofino

- 项目名称：西安卢卡小镇联排别墅样板房
- 项目位置：陕西西安
- 设计公司：上海无相室内设计工程有限公司
- 设计师：王兵、谢萍
- 摄影师：张静
- 用材：白色浑水木饰面
- 面积：310 m²

- Project Name: Show Flat of Townhouse, Lucar, Xi'an
- Location: Xi'an, Shanxi
- Design Company: Shanghai Wuxiang Interior Design Engineering Limited Company
- Designer: Wang Bing, Xie Ping
- Photographer: Zhang Jing
- Materials: Veneer
- Area: 310 m²

整个设计风格传承了意大利文化丰富的艺术底蕴，同时又融入了开放、创新的设计思想，将人们带到举世闻名的旅游天堂波托菲诺。从整体到局部，从繁杂到简单，精雕细琢，都给人一丝不苟的印象。一方面保留了材质、色彩的大气风范，可以很强烈地感受传统的历史痕迹与浑厚的文化底蕴，同时又摒弃了过于复杂的肌理和装饰，简化了线条。在明媚的阳光里，轻声交谈、开怀大笑，尽情享受着醇香的咖啡、葡萄酒和地中海美食。每个人的脸上都洋溢着轻松、安宁、满足和幸福的表情，空气里也弥漫着香甜轻松气息。这便是许多人梦想中的天堂！

A space this project is that inherits the rich artistic backdrop of Italy when fusing design ideals of openness and innovation to expose people to Portofino. The holistic, the part, the complicated and the simple, all are stressed with great care to allow for a meticulous impression. When the grander and generosity has been maintained in materials and colors, the historical and cultural traces can be felt so strong while abandoning the complication and complex of texture and decoration with lines simplified. Chatting and laughing in sunlight with coffee, wine and Mediterranean cuisine. All are easy, peaceful, satisfied and happy. Everywhere is sweet and relaxation. That is the paradise people have in their dream.

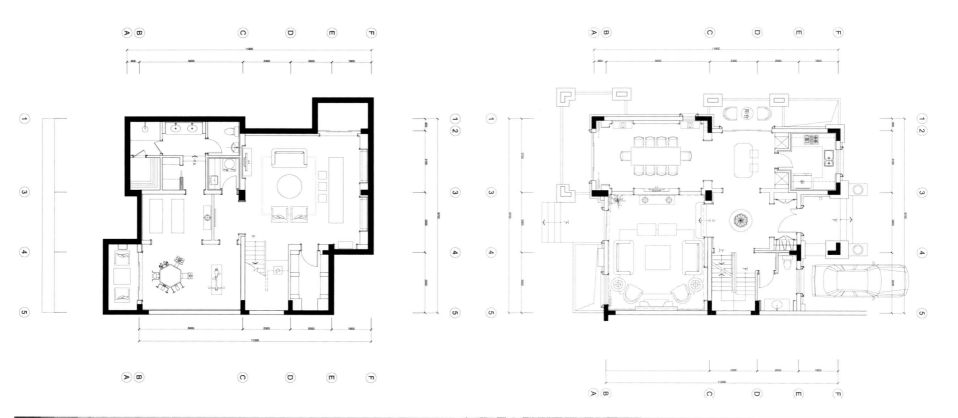

R 清新优雅 柔和静美
efreshing, Elegant and Gentle

- 项目名称：上林西苑叠拼别墅样板房
- 设计公司：上海无相室内设计工程有限公司
- 设计师：王兵、王建
- 摄影师：张静
- 用材：雅士白、古堡灰大理石、彩色烤漆、墙纸、马赛克
- 面积：250 m²

- Project Name: Villa Show Flat, Lucar, West Garden
- Design Company: Shanghai Wuxiang Interior Design Engineering Limited Company
- Designer: Wang Bing, Wang Jian
- Photographer: Zhang Jing
- Materials: Marble, Colored Stoving Varnish, Wallpaper, Mosaic
- Area: 250 m²

徜徉在卢浮宫、香榭丽舍大道，丰富的阅历得以对生活的真谛有着充分的理解，此案设计以清新与优雅为格调，注重人文气息的营造，儒雅中贵气涌动。陈设注重区域组景构图，逐级递进，通过陈设加强整个空间的整体性。客厅是整个空间陈设设计最具表现力的功能区域，特别是在家具及陈设品的表现效果上尤为明显：象牙白色的木饰面、丝织物的柔顺光滑、釉面瓷器瓶艺的晶莹剔透等，浅色系的沙发与暗金质地的茶几组合配以浅灰色的真丝地毯、清新明快的抽象油画、典雅的水晶灯具这一切穿越时空展现一幅低调奢华、贵族生活般的精彩画卷。

The wealth of life experience liking walking around Louver and along the Avenue des Champs Elysees helps to contribute to a full understanding of life. This project stresses more a tone of being refreshing and elegant and the creation of culture to join the learned and refined with the rich. Accessories and furnishings are paid to attention to be unified into the whole. Items within make the most expressive element, particularly the furniture and the accessory. The ivory-white veneer, the soft fabric, the transparency of porcelain, and the light-hue sofa and the tea table in dark gold makes a picture with grayish silk carpet, clean and crisp oil painting in their abstract sense, and chandelier, where you seem to go through the tunnel of time and space.

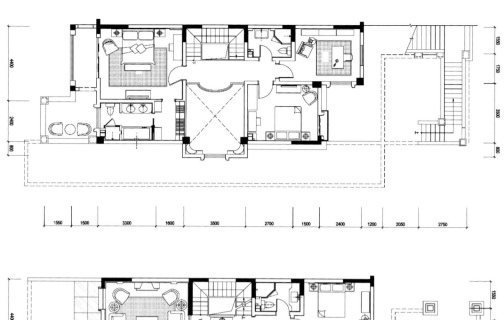
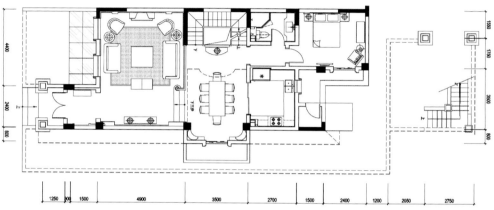
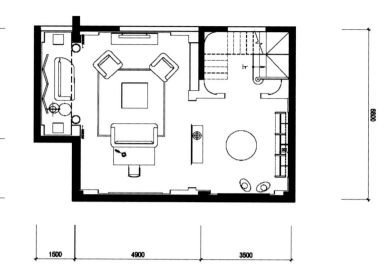

岁月静好 陌上花开
Peace of Time, Flowers in the Field

- 项目名称：荣禾·曲池东岸二期 C 户型
- 项目位置：陕西西安
- 设计师：郑树芬
- 参与设计：杜恒、陈洁纯
- 摄影师：叶景星
- 面积：242 m²

- Project Name: Unit C, Second Phrase, East Bank of Qu Pond
- Location: Xi'an, Shanxi
- Design Company: SCD
- Designer: Simon Chong
- Participant: Amy Du, Jiechun Chen
- Photographer: Ye Jingxing
- Area: 242 m²

法国乡村，一个连接着所有美好记忆的浪漫地域，拥有着最安静、最原生态的生活氛围。在这里可以慢慢品味阳光下摇曳的树叶、石墙上见人乍惊的小鸟和空气中弥漫的悠闲气息。

这里宁静安详，流动着自然清新的气息，自然精致的家，将带你重回岁月静好的质朴生活。

室内空间的布局一气呵成，对称讲究，和谐统一，自然考究的材质肌理与素雅的色调搭配，尽显淡然精致，令视觉上赏心悦目。设计语言中融入原木材质、法式元素，赋予空间淳朴的田园气息与法式浪漫气质。

简约的原木家具、布艺都掩盖不了质朴表象下的品质追求，这些追求是掠过繁华后的素锦，而隐藏在这素锦背后的是清浅的浪漫法式风情。

与沙发同宽的棉麻刺绣布画成为客厅的主角，画面丰富细腻，色调安静温暖，与灰蓝色的地毯相呼应。天然质朴的亚麻色调成为主色调，各个空间的色调既相成一体，又通过一些造型别致、富有质感的原木家具来丰富色彩的层次感。

无论是客厅还是家庭厅，在搭配方面非常讲究，细节之处散发出浪漫的气息。色彩柔和的西洋风情图案地毯，造型优美的法兰西瓷器，棉麻刺绣布艺，花卉床品等都散发着淡淡的优雅，清浅的浪漫气息氤氲在每个角落。浅杏色的肌理明细，手感舒适的刺绣布艺餐椅，与同色系脉络纹理的大理石地板相搭配，成就温馨愉悦的就餐氛围。

主卧亦散发着浪漫的气息，浅色调的玫瑰花壁纸、花卉床品、白色流线形木柜和柔美的陶瓷摆件，如法国少女般温婉。小汽车、棒球、探险等都是小男孩的梦想，天蓝色的儿童房将这些小梦想都收容进去，为他建造一个快乐的生活空间。温和的暖黄色调让次卧充满舒适感，亚黄色的法式花鸟图床品，墙壁上的花卉画都散发着淡淡的优雅。

整个空间延续了法式田园风格浑然天成、清浅浪漫的格调。没有过多的华丽装饰，反倒用自然简约的线条、色彩来营造出田园舒适轻松的氛围。法式软装元素融进简朴的空间，谱唱典雅浪漫的空间曲调。

The country of France makes a place where to join all good memory and romance, to have the most peaceful and the most primitive life atmosphere, to taste at leisure the tree leaves with light cast on and to see the frightening bird and to capture the cozy air.

Here is peace and quiet, where to flow air refresh and nature, and to have a delicate home where to bring you back to tranquility.

Here has a coherent layout, symmetrical and harmonious, where to stress material texture and match of simple hues to set off the detached and the delicacy, to make the vision pleasant, and where to fuse log and French elements to allow for rustic atmosphere and French Romance.

The simple log furniture the fabrics have no way to hide the quality pursuit into the appearance that undertakes no decoration. Behind prosperity is the truth, and the truth is the French Romance.

The cotton and linen of embroidery as wide as the sofa is now the protagonist in the living room, whose picture is rich and delicate, and whose tone calm and warm to echo with the gray blue carpet. The natural and rustic linen now works as the tone. Hues of all sections are integrated while furnishings of log are to enrich the layers with different modeling and texture.

Whether in the living room or in the family room, the match is paid significant attention and the details ooze romance. The western carpet in soft color, the beautiful French china, the cotton and linen of embroidery and the floral bedding are all graceful and element, with romance overspread everywhere.

The apricot texture is delicate and subtle and the chair of embroider feels comfort, making a good match with the marble flooring with the same hue and texture. Both together accomplish a warm and joyous dining atmosphere.

The master bedroom is romantic, where rose wallpaper is of light color, and the floral bedding, the white stream-lined wood cabinet and the ceramic ornaments are like mild French girls. Car, baseball and adventure are dreams of young boy, which are included into the sky-blue child's room for a happy life world. The yellow fills the comfort into the secondary bedroom, where the light yellow French floral bedding and the flowers on the wall are of elegance.

The whole space carries forward the pastoral French style, refreshing and romantic, without overmuch decoration, but more line simplicity and hues to build up ease and comfort. A space it really is, where to accommodate French upholstering elements to highlight a graceful and romantic melody.

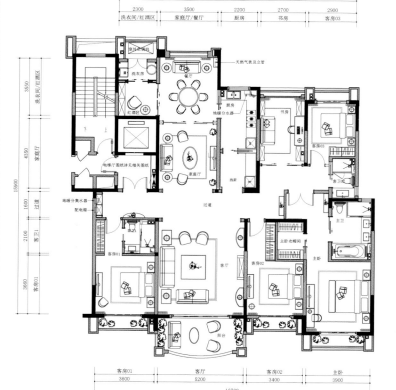

聚焦惊艳新古典
Focus on Neo-Classical Surprise

- 项目名称：风动三泰
- 设计公司：风动设计
- 设计师：何三泰
- 撰文：林雅玲

- Project Name: Peace
- Design Company: PNEUMTIC Internationnal Interior Design Co., Ltd.
- Designer: He Santai
- Text: Lin Yaling

空间设计是一门相当繁复的人文艺术，需要多样化的美学根基，才能将所有的平衡拿捏得恰到好处，尤其是色彩的处理！放眼空间，可用的材质种类琳琅满目，现阶段设计者面对的最大考验，不再是找不到独特材质，而是如何在庞大材质流中，筛选出兼具独特美感与实用性的项目，好为空间创造更棒的气氛与惊艳感。使用整个楼层来规划的主卧室，有着超乎想象的宽敞与独立，设计师何三泰在临窗景观最佳的位置，以架高的地板层次搭配内嵌式的床座样式，营造度假别墅般的尊荣、休闲感。床头墙面选用沉稳的卡其褐色系，衬托造型壁灯的时尚感，床侧精心安排起居厅，焦糖色烤漆玻璃特制的电视墙，提供惬意的视听娱乐。还有窗外的露天花园引人入胜，蓝天绿海环绕的无敌景致，佐以大量温暖木质打造的休憩场域里，导入南太平洋岛屿风情的迷人魅力，遮阳伞、竹编海滩椅、造型卧榻，还有沁凉的马丁尼，共同编织出一段心旷神怡的美丽时光。

As a quite complex literature and art, space design needs diverse aesthetic base to treat with a balance to a very proper point, particularly the treatment for hues. With numerous materials, the real challenge for the designer is not the unique material available, but how to select project with unique project and practicality for a better atmosphere and stunning surprise.

The master bedroom made the best use of the whole floor, is open and separate beyond imagination. The optimal viewing position by the window provides the dignity and leisure only available in resort villa by matching the elevated flooring and the inbuilt bed-like chair. The khaki brown around the bed head sets off the fashion of the wall lamp. The living room is deliberately fixed beside the bed, where the TV backdrop of burnt-sugar color glass offers leisure of audio-visual entertainment. The open-air garden is so appealing, where the scene of blue sky and clear sea is incomparable, which with the resting area of warm wood takes in the charm from the islands in the South Pacific. The sunshade, the bamboo beach chair, and the cool Martini, all together make a joyous and happy time.

S 窗外四季，窗内静水流年
Seasons Outside, Peace Inside

- 项目名称：水景汇实品屋
- 设计公司：夏克设计

- Project Name: Show Flat of Waterscape
- Design Company: Xia Ke Design

对于水景汇实品屋的室内空间规划，设计师以优雅而独特的新古典美学语汇深刻传达出空间品位与内涵张力，让一种优雅、温婉而又高贵的生活品位与格调在空间流淌。

享誉世界的建筑大师安藤忠雄（Anto Tadao）说过，"建筑和自然环境兼容，让生活随时感受环境土地的四季变化，才是好建筑"。因此，在本案空间中，设计将建筑和空间的依存、互动关系转化成一种存于天地之间的永恒对话，以开放、流动、穿透等手法在公共场域甚至是私密空间设立大面开窗，将空间与自然完美融合，让光影水绿在空间中恣意流动、穿梭，打造最舒适的生活尺度，使窗外四季美景成为日常生活中最美的景深，赏景、赏境，更赏心！

整体空间以白色为基调，以简约的石膏线条点缀出欧式空间的特点，加上典雅的舒适家具，整体淡雅而大气，有一种优雅的气质散发其中。客厅主墙面以天然大理石纹理，呼应出空间的绝妙气势与主人独特的生活品位。沙发背景墙则利用清玻璃结合白色镂空工艺设计替代传统实体墙设计，界定起客厅与书房的关系，有效地放大及延伸空间感受。开放式的设计，使餐厅区域与客厅作连贯对应，延伸出敞朗、大方尺度。至于餐厨空间的隔断设计，则与书房和客厅的设计同出一辙。

私密空间通过落地玻璃界面，让空间与自然、光与影结合，而简约的古典语汇，主墙画框式的安排以及整体的婉约色彩，既拉阔空间的大方气势，又让浪漫的气氛盈满一室。

Here makes a project with neo-classical language of unique grace to present spatial taste and meaningful texture, flowing out a life of mild and noble elegance.

As Anto Tadao is quoted, only when compatible with natural for life to experience the seasonal changes, can a building be good. This project consequently transforms the mutual existence and interaction between building and its surroundings into an eternal dialogue between the sky and the land. In public and even private spaces, large windows are fixed to integrate the internal and the external, so that light and shadow, water and greenery can be taken in to make an optimal comfort for life, where seasons find an easy access into life routine. Life is lived in appreciating scenery, and scenery is appreciated in living life.

White on the whole, the space involves simple gypsum lines to shape European traits, which oozes a graceful nature with elegant and comfortable furnishing. All are simple, refined but of grandeur and generosity. The main wall in the living room is of marble to bring out an excellent momentum and personal taste of unique grace. The sofa backdrop of clear glass and white hollowed-out process takes place of the usual wall, defining the boundary of the living room and the study to effectively maximize the spatial experience. The open design links the dining room and the living room, whose partition remains the same as that between the study and the living room. The private space is separated with glass, where to blend nature, light and shadow into the space. Simultaneously, the simple but classical vocabulary, the picture-frame main wall and the mild hues on the whole not only enhance the grandeur and generosity of the space, but injects romance everywhere.

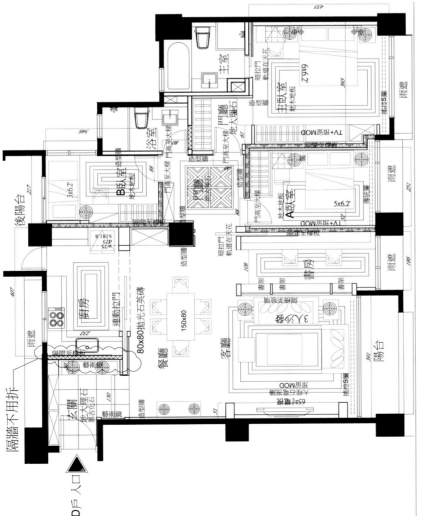

雅奢主张之简美风情
Minimalism of Sumptuous Grace

- 项目名称：荣禾·曲池东岸一期 10 号楼 A1 户型
- 项目位置：陕西西安
- 设计公司：SCD
- 设计师：郑树芬
- 参与设计：杜恒、黄永京
- 撰文：张显梅
- 面积：750 m²

- Project Name: Unit A1, Building Ten, East Bank of Qu Pond
- Location: Xi'an, Shanxi
- Design Company: SCD
- Designer: Simon Chong
- Participant: Amy Du, Jimmy Huang
- Text: Emma Zhang
- Area: 750 m²

曲池东岸 A1 户型整个空间奢雅尽享尊贵，设计师在设计方面以奢、雅、质、暖为关键词，谱写了新一代雅豪们对雅奢生活之道的追求，为当地的住宅翻开奢华雅致的新篇章。

奢：雅豪之士对奢华生活之道的追求
奢乃奢华不落俗套，低调体现文化底蕴而不炫富，是雅豪们对欣赏美，享受美的理解。项目分为三层复式，一层、二层为主要功能区，三层为视听室和棋牌室。其中一层分为客厅、中西餐厅、老人房及客房、休闲厅；二层为主卧、男孩房、女孩房、家庭厅、品茶区。四世同堂、天伦之乐近在咫尺。特别是客厅挑高 7.5 米的中空上层，尊贵大气，主卧及老人房的八角窗极显奢华，完全符合了雅豪之士的品位需求。

雅：触动高贵优雅的心
在设计中，设计师探索对应精神认同层面的需求和设计本身具有的人文关怀，发挥创新的力量，打破传统土豪们的奢侈之念，以全面的顶尖设计手法融入。在空间格调方面，整体清亮有泽，个性的简美设计手法，使空间温暖而高雅，完全体现雅豪们对高品质生活的追求。
休闲厅以高尔夫为主题，球包、书架、奖杯等精致饰品搭配，茶几上饰品的摆放均混合着美式经典韵味，解读了一个雅豪对生活的理解。

质： 设计上的视觉冲击
质乃质感，物体的真实感，丰富多彩做工精细，产生视觉上的冲击效果。设计师以精心挑选的材质及艺术品的搭配表现人本设计理念。家具材质和款式均为郑树芬先生指定的 BAKER 品牌家具，设计师以简洁大气的欧美韵致，选用细腻缜密的布艺、木、金属等，简洁的表象下隐藏着尊贵的内涵，陈设没有多余的造型和装饰，一切皆从功能及舒适着手，从本真出发，使空间整体气质显得更为精致尊贵。
在元素方面，仍以注重简美为主线，除了有软体家私之外，设计师以经典实木家私及实木框架软装沙发做搭配，去掉了繁复的细节，简洁明快大气有形，打造有视觉冲击效果的、极具生活品质的家庭氛围。

暖：春暖花开的日子

暖乃温暖，温馨、舒适之家。家庭厅融入了些许中式韵味，背景墙以叶片形式装饰挂件吸引眼球，局部跳出柠檬黄，充满暖暖的家的味道。八角窗的主卧传达了主人开阔的视野，暗紫色的窗帘，配合着紫红色的晚礼服表达了女主人对生活的追求，从色彩、饰品及主人的喜好无不体现了一种优雅浪漫的生活。

冰雪奇缘、白雪公主、小提琴、照片墙、老唱片、红酒……这一切正昭示着美好生活的开始，永不落幕！

This is a project that enjoys luxury, grace, dignity and honor, which leads to the key words for its description, of luxury, grace, texture and warm to outline the demands of another generation of refined scholars, making new chapter for the local real estate in terms of luxury, brilliance and grace.

Luxury: Pursuit for Life Routine of Refined Scholars

Luxury can be extravagance but should not be chained to the vulgar to embody a cultural setting without showing off wealth. This makes real understanding of and appreciating aesthetics of the refined scholars. Among the three floors, the 1st and 2nd floors work to function, while the 3rd is used for audio-vision room and rooms of chess and card. The 1st houses the living room, the eastern and western dining room, rooms for parents and guests and leisure lobby, while the 2nd the master bedroom, rooms for daughter and son, the family room, and the tea room. With four generations under the same roof, the happiness of family relationships is ready to have. The upper of the living room is elevated 7.5 meters to set off the nobility and magnificence. The bay window in the master bedroom and the bedroom for the parents

bring out an ultimate luxury, completely fitting well in with the tastes of the scholars.

Grace: To Touch Elegant and Noble Heart

During the process, designers explore the corresponding spiritual requirements to make innovation and break away the usual luxury on the basis of the human care inborn with design. Top design approaches are involved simple and unique to boast the spatial warm and grace. The style is bright and glistering on the whole.

The leisure lobby takes golf as its theme, where golf bag, book shelf, trophy and accessories on the tea table interpret the understanding of life of fined scholars.

Texture: To Strike Visual Effect on Design

Texture is the physical existence of object. Items done meticulously are bound to exert strong visual effect. The careful selection of materials and the match of art pieces are human-oriented. Materials and styles of furnishings are of Baker at the request of Simon Chong. Delicate and rigorous materials like fabric, wood and metal are preferably employed with simple appearance implying honorable connotation. As for accessories, there is nothing redundant. All are functional and comfort so the whole temperament is boasted subtle and honorable.

Elements focus on simplicity and aesthetics. Besides the upholstering, classical products of solid wood and solid wood frame are equipped with sofa, leaving off complicated details, simple and compact to exert strong visual effect and shape a

family ambiance of life quality.

Warm: Time When the Weather Is Warm and Flowers Are in Blossom

Warm does not mean so literally, but implies more like something sweet, and here it's also about a comfort family. The family room is added with some Chinese flavor. The backdrop focuses human attention with leaf pendant. The hue of yellow then jumps out of overspread the warm and sweet. The bay window in the master bedroom is the expansive view of the host. The dark purple curtain with the purple night dress is the pursuit of the hostess of life. Hues and accessories all reflect a life of grace and romance. The ice and snow, the snow princess, the violin, the picture wall, the old record and the wine, all are good signs to start good life, which will on no account fade away.

净白美宅，静品时光如歌行板
The Purity of White to Appreciate a Good Time

- 项目名称：贝拉·莫里实品屋
- 设计公司：雅群设计
- 面积：280 m²

- Project Name: Beira Show Flat
- Design Company: Arch Interior Design
- Area: 280 m²

本案280余平方米的实品屋，以白色为底色，采用大量的石材、壁板、镜面元素以及收藏品，呈现出一种大气典雅的人文气度。

客厅、厨房及餐厅等公共空间采用开放式空间规划，餐厅置于客厅、厨房中间，运用弹性。同时，设计也在天花板上做出不同的巧妙设计，界定起其不同的功能空间。客厅布置简约，借由洗练的线条和语汇，成就环境性格和清穆氛围。大量艺术藏品铺饰的沙发背景墙成为了整个客厅场域的主角，展现出艺术人文气势。客厅前有半圆形大阳台，地面铺设48厘米×98厘米施釉抛光石英砖。厨房附人造石台面的"П"字形厨具，并有电器柜、进口三口炉、洗碗机等配备。

主卧室拥有大面采光窗，空间宽敞，规划有更衣室，并附有独立卫浴设备。主卧室同样以白色为主调，为使空间不过于清淡，设计又于空间中适当地加入些紫色、金色和黑色，使空间彰显出低调奢华的美感。主卧卫浴配备台面式双洗面盆、降板式泡汤池、悬壁式马桶，淋浴间内有花洒等。

Beira Show Flat covers an area of over 280 square meters in a white-dominant setting where to employ marble, siding, mirror and collection in large amounts to present a cultural sense of grandeur and generosity.

Public spaces like the dining room, the kitchen and the living room are designed open with the dining room fixed between the living room and the kitchen to allow for flexible use. Ceiling is skillfully defined to divide different spaces. The living room is embellished simple with refined language and lines to shape the spatial personality and the solemn atmosphere. The backdrop of the sofa holds large quantities of art pieces, serving as the protagonist to present the air of art and culture. A large semi-circle terrace in front of the living room is paved with polished glazing quartz bricks measuring 48 centimeters wide and 98 centimeters long. The kitchen counter surface is of artificial stone equipped with kitchen ware in a square form but with the bottom line missing as well as electrical cabinet, imported stove and dish-washing machine.

The master bedroom takes in daylighting in large patches, a section fixed with dressing room and separate bathroom, when dominated with white color to ground off the cool by using extra hues like purple, gold and black in setting off the low-key but luxurious aesthetics. The bathroom is decorated with a double lavatory basin, plate-type bath tub, suspended closestool and an individual shower cubicle.

碧波上的浪漫
Romance Above Waves

■ 项目名称：安缦威尼斯度假村　　　　■ Project Name: Aman Canal Grande Venice

安缦第 25 家度假村位于享有"水上香榭丽舍"美誉的威尼斯大运河之上。安缦度假村所在的圣保罗区是威尼斯六个区域中最小的一个，根据圣保罗教堂而命名，其面积不大却充满神奇的历史色彩，以其各式各样的宫殿、富丽堂皇的教堂和商业繁荣的市集而闻名。安缦威尼斯度假村由两幢相互毗邻的 5 层楼高的建筑构成，共有 24 间套房，其中一个建筑始建于 16 世纪。

乘坐富有威尼斯特色的"水上巴士"贡多拉，沿着威尼斯大运河向着安缦驶去，靠岸后可直通安缦度假村的接待大厅。大厅中光洁如镜的旋转楼梯直通安缦威尼斯度假村最华丽的楼层，包含 3 间风格迥异的餐厅，供应意大利的特色美食，同时也提供种类繁多的亚洲菜品。其酒吧区也占有极好的地理优势，不仅有舒适的设施环境，还有浑然天成的威尼斯大运河美景作为窗边最动人的画卷。

安缦美发馆分成两块区域，均可透过开阔的窗户俯瞰大运河的浪漫景致，风格豪华却不失休闲气氛，客人于此还可享受精美的茶点。Stanza del Tiepolo 则提供国际象棋、西洋双陆棋以及拼图等娱乐设施。Stanza del Guarana 是私人会所的理想场地，这里提供奢华的私人宴会，或安排私密的小型会议，抑或是私人的电影放映。度假村的升降机可直通屋顶露台，这里可以将整个城市的迷人景色尽收眼底，在日落或日出时于此地小憩，更是风景宜人，让人自醉。度假村的花园露台为少有的直面威尼斯大运河的私人花园，在温暖的季节，可以一面以大运河的怡人风景为画，一面以精美餐点为食，仿若在赏画，又宛如置身画中。

安缦度假村共有 24 间套房，并最大限度地保留了过往艺术与建筑风格的壁画与浮雕。大部分套房的客厅与卧室相连接，还有独立的更衣室和浴室。度假村因套房面积的大小、风景视野以建筑风格的不同而分为 4 种套房类型：直面私家花园的豪华套房、可直接观赏威尼斯大运河的运河美景套房、可俯瞰威尼斯大运河全景的行政景观套房以及高级套房。

低矮的房梁，昏暗的灯光，整个水疗室散发出一种避世归隐的氛围。水疗室共有 3 个房间，每个房间均有独立的更衣室及浴室。此外，度假村还有一个小型的健身房。

Sitting above the Canal Grande Venice, a river honored as Elysee above Water, Aman Canal Grande Venice, makes the 25th resort of Aman. Its geographical location is the smallest one of the 6 districts of Venice, St. Paul District named after St. Paul Church and with a historical mystery, where palaces, churches and markets are clustered together. The building of Aman Canal Grande Venice consists of two neighboring constructions, whose 5-floor space accommodates 24 suites. Of the two volumes, one dates back to the 16th century.

By Gondola, a very special vehicle called "water taxi", the journey starts along the river toward Aman and directly into its reception hall, where a mirror-like spiral staircase winds upward onto the most splendid floor, on which there are 3 different restaurants offering Italian specialty and

Asian cuisine. The bar occupied in a premium position not only has comfortable amenity, but also gets a good view of the canal.

Aman Salon is divided into two parts, both of which is waterscape-oriented. In a luxurious but causal atmosphere, guests are served with delicious refreshments. Stanza del Tiepolo features international chess, backgammon, and puzzle. Stanza del Guarana makes an ideal private club, where to have sumptuous private feast, small meetings and movies. The lift leads to the roof terrace, where to get a panoramic skyline of the city. At sun rising hour or sun setting time, you feel nothing but lost in the intoxicating landscape. The garden on top accomplishes rare place to enjoy the river view. The scenery river in warm climate, when accompanied with good meal, completes a picture, where you feel nothing but indulged in it.

All the suites maximize the past art, fresco and relief. With separate dressing room and bathroom, most of the suites are linked with the bedroom. Four types comes based on different size, view and landscape. Beside the deluxe where to taks direct access to the fine view of the river and the panoramic Executive, another is Senior. As for Spa, it has low and short beam and dim light to highlight a reclusive atmosphere. All 3 rooms each are endowed with separate changing room and bathroom. Additionally, there is a small gym right here waiting for the guests.

庄重之美，令人钦佩
Solemn to Be Admirable

- 项目名称：魁北克大公馆
- 项目位置：加拿大魁北克
- 设计公司：大富之家
- 面积：1 385 m²

- *Project Name: Mansion in Quebec*
- *Location: 5 Rue Dublin Candiac, Quebec, Canada*
- *Design Company: Homes of the Rich*
- *Area: 1,385 m²*

"魁北克"大公馆石砌结构，欧洲大陆式灵感空间，建于 2008 年，面积近 1 385 平方米。巨大的空间设有 2 个入口，8 个卧室，7 个浴室，1 个两层的玄关。玄关内一螺旋 Y 形状楼梯婉转直上。除了有 1 个正式的餐厅外，空间另设有 1 个美食厨房和 1 个家庭室。家庭办公室足足占了 2 层空间。另外，空间还有 1 个带有水吧的台球室，还有酒窖、健身房、家庭影院、带有 Spa 的室内泳池，以及可以同时停放 5 辆汽车的车库。室外设有多个阳台，还有带有滑梯的泳池、篝火坑及带有露台的锦鲤池等。

This European-inspired stone mansion is, built in 2008 and features approximately 1,385 square meters of living space with 2 entrances, 8 bedrooms, 7 bathrooms, 2-storey foyer with grand Y-shaped staircase, 2-storey great room, formal dining room, gourmet kitchen, family room, 2-storey home office, playroom, billiards room with wet bar, wine cellar, gym, home theater, indoor swimming pool with Spa, 5-car garage and more. Outdoor features include multiple balconies, patio, swimming pool with slide and swim-up bar, fire pit and a koi pond with gazebo.

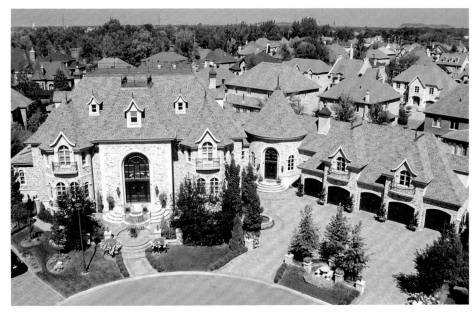

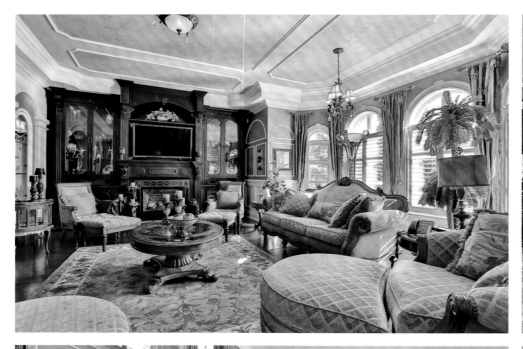

庄园大宅，豪门气度
Manor of Wealthy and Influential Clan

- 项目名称：米西索加大公馆
- 项目位置：加拿大安大略
- 设计公司：大富之家
- 面积：1 674 m²

- Project Name: Stately Mansion in Mississauga
- Location: Ontario, Canada
- Design Company: Homes of the Rich
- Area: 1,674 m²

"米西索加大公馆"面积约1 674平方米，空间富丽堂皇，颇有气势。超大的空间共有7间卧室、11个卫生间、1个带有双楼梯的两层玄关。除了正式的餐厅、客厅外，空间设有1个外高达两层的大房间、2个美食厨房、早餐室、1个家庭室。另外，空间水吧、酒窖、健身房、家庭影院、带有Spa的室内泳池，以及可以同时停放5辆汽车的车库。室外设有多个阳台、汽车旅馆，并有水景。

This stately gated mansion features approximately 1,674 square meters of living space with 7 bedrooms, 11 bathrooms, 2-storey foyer with grand double staircase, formal living & dining rooms, 2-storey great room, 2 gourmet kitchens, breakfast room, family room, wet bar, gym, home theater, indoor swimming pool with Spa, 5-car garage and more. Outdoor features include a motor court, numerous balconies and a water feature.

收藏田园梦想
Collect Countryside Dreams

- 项目名称：青岛龙湖锦璘原著样板间
- 设计公司：深圳市则灵文化艺术有限公司
- 设计师：罗玉立
- 用材：实木、实色环保漆、水曲柳、锈镜、香槟金
- 面积：366 m²

- Project Name: Show Flat of Dragon Lake, Qingdao
- Design Company: Zest Art
- Designer: Luo Yuli
- Materials: Solid Wood, Environment-Friendly Paint, Ashtree, Rustic Mirror, Champagne Gold
- Area: 366 m²

经典美式风格的主要爱好客群为受过良好教育的中产阶层，他们快乐而自信，在居住环境的营造上，追求简单、经典、悠闲而舒畅的生活氛围。经过三十年的改革开放，这样的客群越来越成为我国中青年购房人群的主力，这一风格定位敏锐地捕捉到时代的变迁，呈现出新时代购房人生活方式的变化。

本案以森林色系为主调，营造出低调、轻松的经典美式空间氛围。明媚的阳光，富有生命力的绿植和房间中各种大自然美丽的颜色完美结合，给整个空间带来愉悦、充满活力的生活氛围。宽大舒适的经典美式家具选用柔软、自然的棉麻和呢料，配合仿旧处理的水曲柳材料，线条随意而干练，使室内充满了自然、清新、舒适的气息。让居住在这里的人，感到由衷的舒适以及愉悦。

平面布局基于设计公司的社会学研究完成，注重家庭成员间的相互交流，注重私密空间与开放空间的相互区分。从人的心理以及生活习惯出发，有效地建立起一种温情暖意的家庭氛围。负一层集合了家庭影院、棋牌室和花园等一系列休闲娱乐的生活空间。平面是家居生活功能的承载体，好的平面布局为人的生活提供场所。想象一下在这里，在电视的陪伴下、锅碗瓢盆的和声中、孩子欢快的嬉戏声里，家庭生活其乐融融。

Those with a preference for classical American style are mainly well-educated middle class. They, happy and confident, seek the simple, classical, comfortable and leisure out of the dwelling spaces. With 30 years of reform and opening up, such a target group has been a major force to backbone the purchase of house. Such a style is shrewd to capture changes of the era to meet personal needs and demands of a new period.

The project with a hue of forest color scheme shapes a low-key, relaxing classical American atmosphere, where to impeccably combine sun, greenery and natural colors for presenting a life air of joy, vigor and vitality. The classical American furnishings employ cotton, linen, and wool that with old-treated ashtree of casual and refined lines overspread a sense of being natural and refreshing and comfort, so occupants feel nothing but easy and joyous.

Based on the social study, the layout stresses the interaction among families and accentuates the boundary between the public and the private. A warm family relationship is thus made with human psychology and life habit taken into consideration. The basement houses home theater, chess room, garden and other recreation sections. Only a good layout can be a carrier where people can live in peace and happiness. With sound from TV, cooking and playing children, what a good life here has.

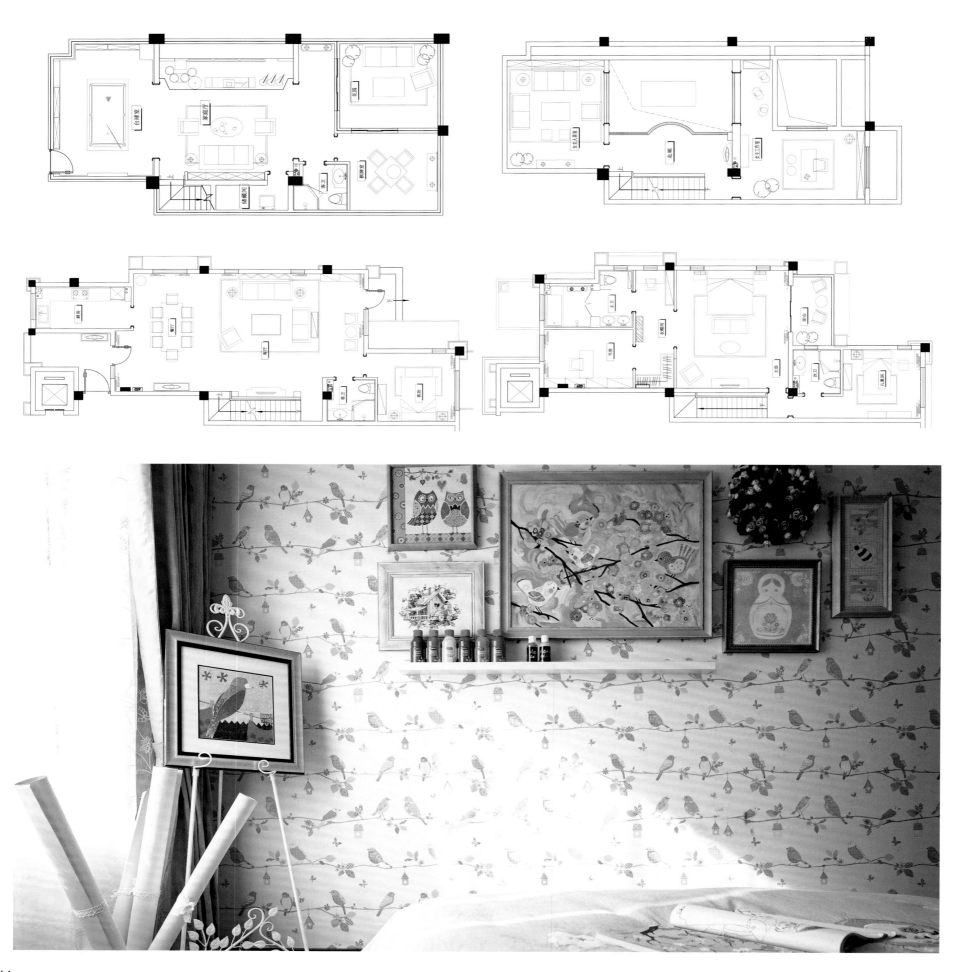

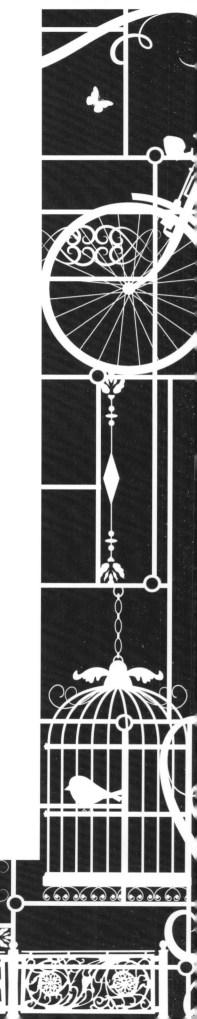

天使的空中城堡
Angel's Castle in the Air

- 项目名称：乌山荣城
- 设计公司：十上设计
- 设计师：陈辉

- Project Name: Glory City, Wu Mountain
- Design Company: Tenup Design
- Designer: Chen Hui

"我想创造一个让人们能在其中慢下来享受生活、纾解压力的环境。推门而至，一种大隐隐于市的氛围将居所静静地包围着，居家的气氛与大自然的气韵在此融为一体，在这样的环境中，家人可以其乐融融地一起交流。"设计师陈辉将悠扬的岁月雕刻在居家生活中，给人以温暖、质朴的感受。

公共空间有别于一览无遗式的开放式设计，多层次的空间布局有着丰富的视觉效果。客厅采用下沉式设计，体量厚重的座椅均采用深沉的色彩，衬托之下就有了落地生根的扎实感。不规则的文化石墙面既是壁炉又承载着空间区隔的作用，无形中家的动线也随之增加。粗糙的文化石、皮质的沙发、金色的豹纹椅，野性与奢华的自由组合愈发让家散发出时尚的魅力。

客厅与餐厅之间虽有隔断，依旧处在同一个空间范畴里，设计师利用同色系的吊顶、同风格的灯饰和共用的落地玻璃顺利过渡，使家居氛围更为和谐统一。岛台设计通常是安排在厨房之内，本案却外露在开放式的餐厅内。在看过狭长的厨房后，就会发现如此设计不仅避开了空间的局限，还增加了吧台的功能。仿古的地砖搭配传统的美式家具，柔和的色彩给人以轻松、温馨的家居氛围。

真正的家不仅仅是一个展示的空间，它还是一个足以容纳爱的空间，而这份爱就体现在细节。在起承转合的空间里，家的温馨可以是一面墙画、一束灯光，也可以是一件传承已久的老式摆件。在昏暗的光线下，似乎能听到长辈细语、小孩嬉笑的声音，其乐融融。

无论是木饰墙面的主卧还是花鸟图的墙面、刷墙漆的次卧，都有着浓浓的美式味道，历久弥新的风格仿佛岁月的缓慢流淌。在延续这番浪漫情愫的同时，儿童房的新元素则让这个家变得活泼跳跃。设计师因地制宜，利用阁楼的不规则形态为纯真的孩童打造出梦想的海洋。幽蓝的马赛克卫浴在上演着海底总动员，层层叠加的巨轮上装载着小船长游历世界的航海梦想。主题式的卧房不仅满足了孩子自由好动的空间需求，还让天真烂漫的想象有了更广阔的海洋。

This project is met for a creation, where people can be at leisure to enjoy life in a relaxing and pressure-releasing setting, immersed in an atmosphere reclusive in metropolis with life and nature fused together. Time of peace and quiet is lost in home life to allow for warm and plain experience, with which families can live in joy.

Quite different from the usual open design, the public space features hierarchy of rich visual effect. The living room is of sinking design, where the historical chairs are of dark color to set off the sense of being rooted into the land, and wall of irregular cultural pebbles serves also as

fireplace while partitioning the space to add more lines inadvertently. The rough cultural stone, the leather sofa, and the golden leopard-spotted chair bring out the fashion charm with wild but sumptuous combination.

Though separated, the living room and the dining room are fixed under the same roof, whose suspended ceilings are of the same hue, lighting fixtures are of the same style and the landing glass is shareable, all to integrate the dwelling space into harmony and unification. The island is fixed into the open dining room as unexpected, and the narrow kitchen is not cramped any longer but makes room for bar counter. The antique brick with the traditional American furniture provides an easy and warm atmosphere.

A home space in its real sense should not just display, but be more tolerate to house love, which is embodied in details and in a space, where warmth can be a wall coated in picture, a bunch of light and an old accessory that has passed down from one generation to another. With dim light, sounds by elders or playing children are bound to make notes of harmony.

Whether the wall decorated with wood and the wall with flowers and birds in the master bedroom, or the wall wrapped in paint in the secondary bedroom is of strong American flavor, where the newly-refreshed style seems to have been aired. When continuing the romance, new elements in children's room make the home lovely and vivid. The irregular attic is made the best use of to build up an ocean of dream for innocent children. The indigo mosaic in the bathroom makes scenes of Finding Nemo, where huge ships one after another accommodates the navigation dream of the young captain to travel around the world. The thematic bedroom not only meets the needs of children for space, but provides enough space for naive imagination.

复刻慢时光
Slow Time Once More

- 项目位置：中国重庆
- 硬装设计：庞一飞、殷正莉
- 软装设计：夏婷婷
- 用材：水曲柳、实木地板、皮革软包、油蜡皮沙发、陶瓷手工马赛克
- 面积：265 m²

- Location: Chongqing, China
- Decoration Designer: Pang Yifei, Yin Zhengli
- Upholstering Design: Xia Tingting
- Materials: Ashtree, Solid Flooring, Leather, Sofa, Mosaic
- Area: 265 m²

复刻慢时光，是设计师的一种浪漫。正如小仲马所说，我认为只有深刻地研究过人，才能创造出人物，如同只有认真地学习了一种语言才能讲它一样。作家运用文字来描写古今，设计师运用不同的材质撰写小说。本案以自然、优雅、含蓄作为空间关键词，希望空间能够承载更多情感。

设计师认为大多数人这样度过一生好像欠缺点什么。大家承认这种生活的社会价值，也看到了它的井然有序的幸福，但是在品辰设计师的血液里却有一种强烈的愿望，渴望一种更狂放不羁的旅途，渴望一种更加不同的生活。对咖啡色的灵活运用，在空间中营造出未来怀旧的质感。铆钉与皮革的组合是英伦摇滚的细节体现，是对自由的低吟。卫生间的陶瓷手工马赛克，为小空间增添丰富元素。

品辰想表达的空间情感如维多利亚时期诗人丁尼生所写："及时采撷你的花蕾／旧时光一去不回／今天尚在微笑的花朵／明天便在风中枯萎。"

Slow Time once more is actually a kind of romance from designer. Personally, I think only with a good sight into human beings, can people like Alexandre Dumas Fils make creation of figures, just like only the study of a language can lead to proficiency of it. What writers depict the past, the present and the future is to what designers present an interior project with the use of materials. And this project takes "natural, graceful and reserved" as its key words with an aim that the space can be a carrier of more emotions.

According to designers, most people seem to be short of something if they continue to live like this. When acknowledging the social value of such life, people get an eyesight of its happiness that comes orderly. However, this is not the same case of designers responsible for this space, for they have a stronger designer to start a journey that nothing can hinder to fix barrier upward and to have a different life.

The flexibility of using coffee brings out a nostalgia texture that is future-oriented. The combination of rivets and leather is the detailed expression of British rock music, or croon for freedom. The hand-made porcelain mosaic adds more elements to the space.

That is what's destined to be accomplished by the designers, just as Tennyson, a poet in Victorian era writes that, timely pick up your flowers, or time gone would never return, even flowers are still in blossom today are bound to die away in wind tomorrow.

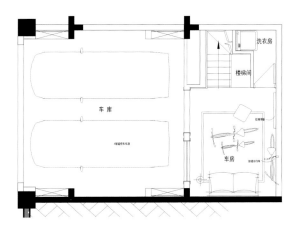
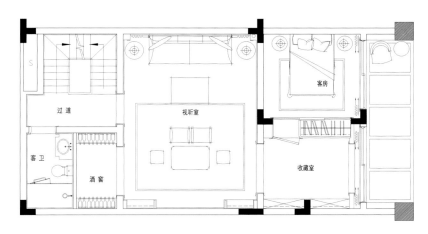
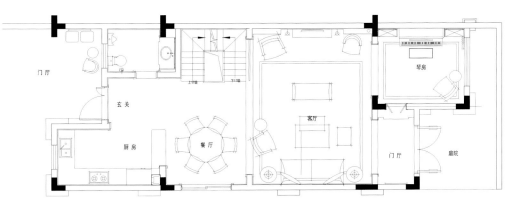
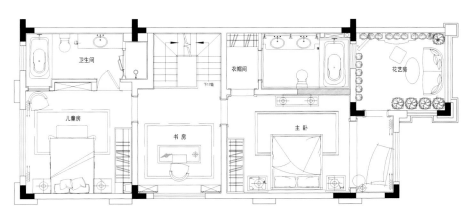

意法经典，都市沙龙
Italian and French Classic City Salon

- 项目名称：意法经典
- 项目位置：加州比弗利山庄
- 建筑公司：兰德里设计
- 设计公司：琼建工
- 面积：2 046 m²

- Project Name: Classical Italian Meets French Art Deco
- Location: Beverly Hills, CA
- Construction Company: Landry Design Group, Inc.
- Design Company: Joan Behnke & Associates, Inc.
- Area: 2,046 m²

"意法经典"位于比弗利山庄极其难得的一个地段。业主希望能有一个集尖端、温暖于一体的独一无二的空间。建筑的意大利式的立面配上了法式的装潢工艺，成就了一个经典。每天100位员工同时作业，7个卧室的公馆终于在2 046平方米的基地上走到了世人面前。

玄关的扇形墙，二楼及开放式的中庭，由来自纽约的娴熟技师倾心打造。所有的窗户都来自于意大利里米尼。从中庭至玄关，到主卧，每处装饰都极其时尚别致。主套开向客厅，面朝卧室以及男、女主人公卫生间。来自托斯卡纳海岸的石块运用于所剩不多的空间内，如主人客厅的核心部位，满足了主人的特别爱好。

娱乐室的精心考究是对业主大家庭生活的满足。宽大的餐厅举行正餐的同时，还可看外面的景观。宽大的厨房精心考究，位于非正式早餐厅的旁边。合适的花岗石台面提升着女主人的烹饪体验。附近的家庭室宁静、舒适，除了休息大堂，还有内置的水族缸。

楼下有室内泳池、热水浴缸、按摩房、桑拿。游戏房模拟着20世纪30年代海洋驳轮的场景。枝形吊灯、钢构天花、主墙面定制的壁画给人一种热闹的船上画面。除了家庭影院，楼下还有可以容纳1 500瓶酒的酒窖。室外，泳池自是不可或缺，小小的户外休息大厅带有烧火坑。从此可以俯瞰泳池及建筑后面的优雅立面。

On a rare piece of real estate in Beverly Hills, these homeowners were dedicated to having a sophisticated, one-of-a-kind home to fill with family and warmth—and this is exactly what they got. The Italian architecture and facade of the building is juxtaposed beautifully with a unique French Art Deco interior. Staffing 100 workers a day, we built this seven-bedroom, 2,046-square-meter residence from the ground up in just three years.

We hired a skilled plaster atelier from New York to create the custom-scalloped walls of the entry foyer and the architectural detailing of the second-floor, open-air atrium. The endless windows throughout the residence were brought in from Rimini, Italy. Across from the atrium is the entrance to the master suite, which opens into an elegant sitting room, bedroom and his-and-her master bathrooms, each clad with sultry, smart antique decor. The homeowners fell in love with an Italian marble that we wanted to use on more of the house, but only a limited quantity was left in existence. Chasing the remaining blocks of stone took a few trips to the Tuscan coast. Finally, when on a cruise with my family not far from the area the stone came from, I had to leave the boat a couple days for one last look. During the brief excursion I found and purchased the remaining blocks of marble and they fit perfectly as a core element in the hearth of the master sitting room.

As one of the most important uses for this home was to gather the homeowners' large family, we worked closely alongside the designer and architect to create the finest entertainment areas. A large dining room for formal meals overlooks the front landscape, while a spacious and meticulously planned kitchen sits adjacent to an informal breakfast room. To be sure the homeowner

had the best experience of their love of culinary arts, I traveled to Italy with the interior designer for the perfect kosher granite counters for the kitchen. The family room nearby offers a serene comfort with a lounge and built-in aquarium.

The activities continue downstairs with a spa containing an indoor pool, hot tub, massage suite, and sauna. For more enlivening gatherings you're led to the one-of-a-kind game room, designed to emulate the deck of a 1930's ocean-liner. Complete with rope chandeliers, a steel-framed ceiling, and a unique trompe l'oeil custom-painted mural across the main wall, giving the genuine ambiance of a ship's deck. Also downstairs, a home theater and 1,500 bottle wine cellar also make for wonderful entertainment. Outside, we built the swimming pool and added a two-story guest house with a small outdoor lounge with a fire pit that overlooks the pool and onto the elegant rear facade of the main house.

醇雅诺丁汉
Nottingham of Fashion and Elegance

- 项目名称：长春大禹样板房
- 设计公司：PINKI品伊创意集团 & 美国IARI刘卫军设计师事务所
- 设计师：刘卫军
- 参与设计：梁义、袁朝贵
- 软装设计：PINKI品伊创意集团 & 知本家陈设艺术机构
- 用材：大理石、木饰面、墙纸
- 面积：230 ㎡

- Project Name: Da Yu Show Flat, Changchun
- Design Company: PINKI & IARI Design
- Designer: Liu Weijun
- Participant: Liang Yi, Yuan Chaogui
- Upholstering Design: PINKI
- Materials: Marble, Veneer, Wallpaper
- Area: 230 ㎡

家，是一种文化；家，是一段时光；家，也是一种情怀。在激情之处，邂逅心中的"美神"，纯净、高贵、典雅……仿佛触不可及，却又身在其中，那是一场午夜时尚的梦。

明明是淡淡的空间，却因为设计师的参与而被渲染成一幅色彩最浓烈、色调最醇厚的美丽画卷，浸染出馥郁的华彩，流泻出醉人的芬芳，让人在沉醉的清醒中，欲醉未醉，迸发出对生活的爱与激情。

张扬的红色调是底色，它鲜明、强烈、奔放，散发着低调而沉稳的气质，带着高贵的质感与醇香，似乎是从时光沉淀的表情中传承新的气质。而充满魅惑的紫、深邃的蓝、时尚经典的黑、高贵典雅的金、含蓄低调的墨绿……跳跃着点缀其中，丰富而饱满，强化了画面的感染力，势将至浓至纯的空间情感推至极致。而家具，这个在设计中扮演着一个十分重要的角色的设计语汇，更是以浑厚简洁的造型展露出时尚复古、典雅精致的气魄，就像老朋友似的，让人体会到亲切而真挚的感受。精致的生活，怎么能缺少花之艺术呢？它是装点生活的艺术，是经过设计师构思、制作而创造出来的艺术品，强烈的装饰性渲染了空间的艺术气氛，画龙点睛地传递出浪漫的情感与生活的情趣。

犹如绵香的醇酒，本案既有新酿的浓烈，又有陈年的沉郁，这种馥郁的芬芳随着时光的沉淀更将留下亲切而沉静的韵味，经久不散，直至永藏于心底，浸入到骨髓，飘落到梦中，铭刻在岁月的门楣。而恰恰又是岁月的洗礼与沉淀，让空间与情感完美地融于一体，触景生情，如滚滚热源一般，提升了空间温度，打动人心。

Home makes a culture, a time and an emotion. In the inner depth of tranquility, home offers a space where to meet with God of Aesthetics, pure, noble and elegant to come to you beyond expectation, but simultaneously you feel nothing but lost and indulged within, feeling as if you were in a dream at midnight.

A space originally simple and light is stunningly rendered into a picture scroll coated in heaviest hues to bring out the intoxicating brilliance and fragrance out of pigment, where people keep drunk and equally sober to have a passion and enthusiasm for life. The dominant color of red, bright, strong, bold and unconstrained, is reserved, staid and calm with a noble texture and a mellow sense. All seem to have got inheriting nature with time gone by. Meanwhile, other colors are interspersed, like purple, blue, black, gold and the very green, abundant and solid to strengthen the influence of the picture in reaching the spatial purity into an utmost level. Furnishings and accessories, as a pivotal role in design, embody a retro fashion and elegant delicacy, like an old friend to make people feel sincere, friendly and amiable. Flowers are then indispensable, as an art to beautify life. As for the floral element available in this project, all deserve wisdom, intelligence, ingenuity and skills, rendering the spatial artistic atmosphere to convey the romantic feelings and life taste like the finishing point.

A space this project really accomplishes that is like a mellow wine with both the strong of the newly brewed and the mature, whose essence left out when time goes by is cordial and tranquil, everlasting and permeable into bones, which you miss day and night even vicissitudes has found its way onto the door header. Space and emotion has fused impeccably and perfectly, feelings stirred up out of sight like heat sources to raise the temperature to move and touch heart and soul.

新手法讲述美国老故事
New Way to Tell the Old American Story

- 项目名称：荣禾·曲池东岸二期4号楼D户型
- 项目位置：陕西西安
- 设计师：郑树芬
- 参与设计：杜恒、黄永京
- 撰文：张显梅
- 面积：280 m²

- Project Name: Unit D, Building Four, East Bank of Qu Pond
- Location: Xi'an, Shanxi
- Designer: Simon Chong
- Participant: Amy Du, Jimmy Huang
- Text: Emma Zhang
- Area: 280 m²

荣禾·曲池东岸荣踞曲江核心，拥享繁华世界，以高雅曼妙的姿态屹立在曲江池东岸，为西安住宅领域开拓出新的想象空间，重构中国历史古都雅豪圈层住区新秩序。其由香港著名设计师郑树芬（Simon Chong）主创设计，郑树芬将其标志性的风格带入项目，即"雅、奢"，结合美式古典与现代品位，保留了殖民地风格中最具代表性的元素特征，着眼当代人居住观念和生活诉求，挑战前沿风格模式，打造出充满艺术和文化的美式风格气息场所。

空间层次丰富，以成熟尊贵的胡桃木色、冷静高贵的高级灰色为色彩基调、搭配安详典雅的墨绿色及热情灵动的红色诉说着主人的低调与内心的高贵。客厅中，色彩厚重的油画、古铜色的吊灯、怀旧仿古艺术品、胡桃木色酒柜、茶几、拼花单椅、壁炉上的古董收藏摆设，简洁而怀旧、实用而舒适，不经意中成就了一种轻松舒适高贵的灵魂。

项目面积近300平方米，分为客厅、家庭厅、餐厅及红酒区、书房。卧室除主卧之外，另外分别为老人房、女孩房及客房。每一个空间都是极恰到好处，郑树芬讲究的是如何通过生活经历去实现自己对艺术的启发及品位的提高，从中摸索出独一无二的美学空间。

美国是一个移民文化为主的国家，它有着欧罗巴的奢侈与贵气，但又结合了美洲大陆这块水土的不羁，设计师郑树芬将主人翁定格为喜欢旅行摄影，书房地毯也营造一种旅行者的感觉，书桌上凌而不乱放着裁切的照片，墙壁上以美国老照片诉说着许多不同的故事。设计师意在以寻找文化根基的新的怀旧方式，迎合时下文化资产者对生活方式的需求，去挖掘一种有文化、贵气、拥有自在与情调感的气质。

家私多种形式呈现，或实木配皮质或拼花或绒布或豹纹，或现代或古典或野性……没有太多修饰语约束，自由随意，造就了自在、豪放不羁的生活方式。

在这里，就如同常见的美国电影画面，有家人的照片在角落里，有不舍得放弃的阳台小花园，有开放厨房绕着全家的笑声，有明亮的浴室让人驱除疲倦。美式居家说的不只是一种风格，也是一种生活态度，相信您也会爱上它。

The compound where this project overlooks the world prosperity and makes a pioneering imagination in lithe and graceful approaches, reorganizing orders in an ancient capital. Into the project are implanted with personal identification of the designer, elegant and luxurious to combine classical American style and modern taste while maintaining elements and traits, representative of Colonials time. The demand to meet local residence and life builds up a field filled with art and culture.

The spatial layers are rich. The walnut color and gray serve as the tone with very green and red to tell of personal low key and inner dignity. The sofa backdrop in the parlor is of oil painting, copper-colored chandelier, antique art pieces and other items like walnut-hued wine cabinet, tea table, parquet chair and antiques on the fireplace. All are simple, reminiscent, practical and comfortable to make a soul of honorable, relaxing and easy.

The area nearly of 300 square meters is divided into parlor, family room, dining room, study and

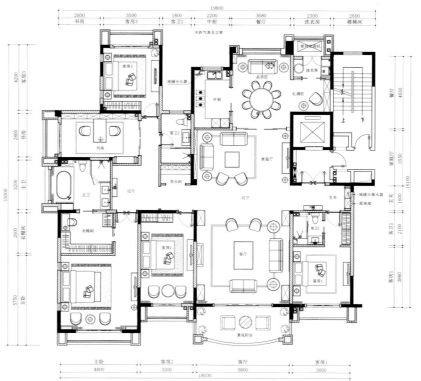

red wine sector, apart from master bedroom and rooms for parents, daughter. Each is treated to the point, for the designer Simon is a person who has accumulated individual spatial aesthetics from life and understanding toward art.

American culture, a diverse pot to melt exotic branches, has luxury and honor of Europe when keeping freedom exclusive to its own. The owner is supposed to be a traveler, so even the carpet in the study allows for a journey feeling. Pictures on the table seems to be put at random and these on the wall tell of stores. The aim to seek new ways to keep memory on the basis of culture caters for the current tread to explore a culture, fortune and personal indulgent life.

Furnishings appear differentiated: some solid wood is equipped with leather, parquet, lint and leopard print, while some are modern and some wild. The freedom without overmuch decoration makes life here as freely as you want.

Here presents scenes usually in American films, where family albums stand in corner, balcony and small garden are always there, and laughter bursts out from open kitchen and bathroom is bound to keep off fatigue. American style is more an attitude toward life than a style. Trust you to live it.

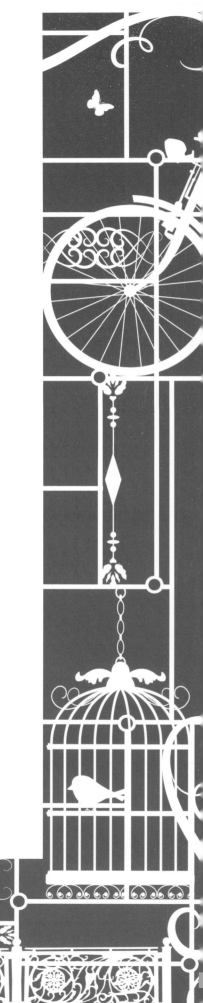

360度环景大宅，山谷里的石头居
360 Degree Panorama Stone Mansion

- 项目名称：迈耶大公馆
- 项目位置：亚历桑那
- 设计公司：罗威设计
- 设计师：马克·罗威
- 摄影：雷摄影
- 面积：465 m²

- Project Name: Meier Residence
- Location: Canyon Pass – Dove Mountain, Marana, Arizona
- Design Company: Soloway Designs Inc.
- Designer: Marc A. Soloway
- Photography: Ray Albright Photography, Tucson, AZ
- Area: 465 m²

"迈耶大公馆"位于山谷之中。陡峭的花岗岩石壁给设计带来极大的挑战，但同时也提供了很多可以实践生态理念的机会，创造出可以360度观景的个案。于是，淹没于自然的轮廓中，建筑成了岩壁的一分子。太阳与光线透过落地窗，墙体窗于室内空间中自由地舞蹈。

南面与西面成了生态的实践基地。内外之间的风景给人一种和谐的生活体验。玄关处，高耸的手工木质、钢松三面立体桶式壁炉别样耀眼，同时也是整个空间的视觉点，整合着设计阴阳的五行观。同样的理念透过3米的瀑布清晰可见。瀑布从窗户直至壁炉，整个水景造型借助于深掘基地的岩石而成，如同天然。

无论室里、室外，还是里外之间，平面布局以最大化地娱乐客人为根本，或予人放松，或提供令人叹为观止的落日美景。

Situated within a canyon, the steep granite walls present exciting building challenges, offer unique ecological opportunities, and conspire to create spectacular 360 degree views. Gently wrapping around natural contour lines, the building becomes one with the canyon walls. Sun and light are pulled into the contemporary, open-space layout through ceiling-to-floor, multi-stacking window walls.

South and west exposures provide eco-friendly opportunities. Views abound inside and out, creating a synergistic living experience. Highlighting the entry is a towering, hand-carved wood and steel, three-sided barreled fireplace. Visible from every part of the main living areas, this artistic centerpiece serves to unify the design's five elements: earth, fire, metal, wood and water. The latter is embodied in the 3-meter waterfall visible through the window to the right of the fireplace, and created with boulders excavated from the building site.

Whether inside, outside, or a combination of both, the floor plan encourages entertaining guests, or simply relaxing and enjoying the panorama of unbelievable sunsets.

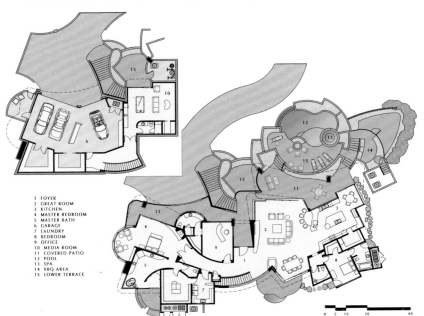

1 FOYER
2 GREAT ROOM
3 KITCHEN
4 MASTER BEDROOM
5 MASTER BATH
6 GARAGE
7 LAUNDRY
8 BEDROOM
9 OFFICE
10 MEDIA ROOM
11 COVERED PATIO
12 POOL
13 SPA
14 BBQ AREA
15 LOWER TERRACE

图书在版编目（CIP）数据

浪漫新古典 V / 黄滢，马勇 主编 . – 武汉：华中科技大学出版社，2015.5
ISBN 978-7-5680-0914-0

Ⅰ . ①浪… Ⅱ . ①黄… ②马… Ⅲ . ①住宅 – 室内装饰设计 – 图集 Ⅳ . ① TU241-64

中国版本图书馆 CIP 数据核字（2015）第 120132 号

浪漫新古典 V

黄滢 马勇 主编

出版发行：华中科技大学出版社（中国·武汉）
地　　址：武汉市武昌珞喻路1037号（邮编：430074）
出 版 人：阮海洪

责任编辑：熊纯	责任监印：张贵君
责任校对：岑千秀	装帧设计：筑美空间

印　　刷：中华商务联合印刷（广东）有限公司
开　　本：889 mm × 1194 mm　1/12
印　　张：24
字　　数：144千字
版　　次：2015年8月第1版 第1次印刷
定　　价：358.00元（USD 71.99）

投稿热线：（020）36218949　　duanyy@hustp.com
本书若有印装质量问题，请向出版社营销中心调换
全国免费服务热线：400-6679-118 竭诚为您服务
版权所有　侵权必究